T0135630

True-amplitude Kirchhoff migration: analytical and geometrical considerations

---

Amplitudenbewahrende Kirchhoff-Migration: analytische und geometrische Betrachtungen

Zur Erlangung des akademischen Grades eines

DOKTORS DER NATURWISSENSCHAFTEN

von der Fakultät für Physik der

Universität Karlsruhe (TH)

genehmigte

DISSERTATION

von

Dipl.-Geophys. Thomas Hertweck

aus

Baden-Baden

Tag der mündlichen Prüfung: 6. Februar 2004

Referent: Prof. Dr. Peter Hubral

Korreferent: Prof. Dr. Friedemann Wenzel

Bibliografische Information Der Deutschen Bibliothek

Die Deutsche Bibliothek verzeichnet diese Publikation in der Deutschen Nationalbibliografie; detaillierte bibliografische Daten sind im Internet über http://dnb.ddb.de abrufbar.

ISBN 3-8325-0512-1

Logos Verlag Berlin
Comeniushof, Gubener Str. 47,
10243 Berlin
Tel.: +49 030 42 85 10 90
Fax: +49 030 42 85 10 92
INTERNET: http://www.logos-verlag.de

When your only tool is a hammer, everything looks like a nail.

Abstract

An important task in seismic reflection imaging is to estimate the subsurface structures from the prestack data. This means that signals in the time domain, the reflection events in the recorded data, have to be transformed into their images in the depth domain, the reflectors. The corresponding process is usually called migration. A geometrically appealing approach for this task is Kirchhoff migration which is based on an integral solution of the wave equation. Applied in its original purely kinematic form called diffraction-stack migration, this process provides a structural image of the target region under investigation. However, Kirchhoff migration is also able to handle the dynamic (i. e., amplitude-related) aspects of wave propagation, thus allowing to assign physically sound amplitude values to reflector images. In such a true-amplitude migration, the geometrical spreading effects of wave propagation are removed from the input data during the imaging process. As a consequence, reflector amplitudes become basically a measure of the angle-dependent plane-wave reflection coefficient. Due to the inherent redundancy of seismic multicoverage data, this allows to analyze the amplitude variation with reflection angle, which provides information about the subsurface that goes far beyond a purely structural image: the changes of medium properties at interfaces can be inferred. This may be helpful for the identification and classification of, e. g., hydrocarbon reservoirs.

My aim in this thesis is to relate the strict mathematical derivation of true-amplitude Kirchhoff migration to clear geometrical concepts. Commencing with the basics of wave propagation and ray theory, a complete description of Kirchhoff migration, the adjoint operation to Kirchhoff modeling, is given. I close the gap between the originally graphical migration schemes and the nowadays available analytical descriptions based on a stationary-phase evaluation of the migration integral. Further aspects relevant to the correct recovery of amplitudes in depth migration, such as the handling of topography and irregular acquisition geometries, are explained in detail in a mathematical as well as in a geometrical manner. All above-mentioned aspects are verified for a synthetic dataset, both qualitatively and quantitatively, confirming that the proposed solutions are appropriate in the context of true-amplitude imaging.

Finally, Kirchhoff migration is integrated into a seismic reflection imaging workflow based on the data-driven common-reflection-surface stack method. This approach allows a consistent and largely automated seismic data processing from the preprocessed data in the time domain to the image in the depth domain. Its application to a recently acquired real dataset demonstrates the potential of this approach, especially with respect to the interpretation of the final migrated image concerning fractures and faults.

It turns out that relating the mathematics of seismic imaging methods to the corresponding geometrical properties is extremely helpful: on the one hand, it allows to gain an intuitive understanding of these methods, their features and pitfalls. On the other hand, it helps to evaluate the quality and reliability of the results prior to their final interpretation.

Zusammenfassung

Vorbemerkung: Die vorliegende Dissertation ist bis auf diese Zusammenfassung in englischer Sprache geschrieben. Da auch in der deutschen Sprache einige englische Fachausdrücke gebräuchlich sind, wurde bei diesen Ausdrücken auf eine Übersetzung verzichtet. Sie werden, mit Ausnahme ihrer groß geschriebenen Abkürzungen, in *kursiver* Schrift dargestellt.

Kapitel 1: Einleitung

Die Erde ist seit Menschengedenken immer ein faszinierendes Untersuchungsobjekt gewesen. Die Ursache vieler Phänomene, z. B. das Auftreten eines Erdmagnetfeldes oder die Verschiebungen der Kontinentalplatten, liegt verborgen im Innern der Erde. Leider ist ein Großteil dieses Bereiches weder für Menschen noch Maschinen direkt zugänglich, sodass Wissenschaftler auf Messungen auf oder außerhalb der Erdoberfläche angewiesen sind, um das Innere der Erde zu erkunden. Geophysiker haben hierzu verschiedene Verfahren und Auswertemethoden entwickelt, wobei einige auf passiven Messungen beruhen, andere wiederum aktive Quellen zum Einsatz bringen. Zu letzteren Methoden zählt unter anderem die Reflexionsseismik, auf die im Folgenden näher eingegangen wird. Bei einem reflexionsseismischen Experiment werden durch eine geeignete Quelle (z. B. ein Vorschlaghammer, eine Explosion oder eine *air gun*) an vielen Orten elastische (oder akustische) Wellen in die Erde gesandt. Die Antwortsignale der Erde werden dann wiederum an vielen unterschiedlichen Positionen mit Hilfe von Geophonen oder Hydrophonen aufgezeichnet. Das so aufgezeichnete Wellenfeld hat einen gewissen Teil der Erde durchlaufen und beinhaltet damit Informationen, mit deren Hilfe sich ein Bild über diesen Teil des Untergrundes erstellen lässt.

Innerhalb des Erdkörpers gibt es im Prinzip zwei Arten wie sich die physikalischen Eigenschaften räumlich ändern: Einerseits gibt es kontinuierliche, glatte Änderungen der Materialeigenschaften, andererseits aber (hauptsächlich in vertikaler Richtung) abrupte Änderungen, auf die die Reflexionsseismik sensitiv ist. Diese scharfen Diskontinuitäten gilt es im Laufe der Verarbeitung reflexionsseismischer Daten zu finden und abzubilden.

Das Verfahren, das die im Feld gewonnenen und später vorverarbeiteten Daten vom Zeit- in den Tiefenbereich transformiert, wird Migration genannt. Die Aufgabe der Migration ist es, die Position und die Neigung von Reflexionsereignissen korrekt darzustellen, sodass ein Bild entsteht, das möglichst genau die Strukturen im Untergrund widerspiegelt. Zudem muss die Migration Diffraktionen und Triplikationen auflösen sowie den Einfluss der Topographie berücksichtigen, falls die Messdaten nicht auf einer horizontalen Ebene gewonnen wurden. Bei einer amplitudenbewahrenden (*true-amplitude*) Migration kommt ein weiterer Punkt hinzu: der Effekt der sphärischen Divergenz (*geometrical spreading*) wird korrigiert. Dieser Effekt ergibt sich dadurch, dass sich die Energie in Raumwellen auf eine in der Regel immer größer werdende Wellenfront verteilt und somit die Energiedichte abnimmt.[1] Wird der Effekt während der Migration rückgängig gemacht, so sind die Amplituden im migrierten Abbild ein Maß für den winkelabhängigen Reflexionskoeffizienten. Auf diese Art und Weise erhält man nach einer amplitudenbewahrenden Migration also nicht nur ein strukturelles Abbild des Erduntergrundes, sondern auch quantitative Informationen über die Änderung physikalischer Eigenschaften an Schichtgrenzen. Das stellt die Grundlage für so genannte *amplitude-variation-with-angle* (AVA) Studien dar, mit deren Hilfe sich z. B. Öl- oder Gasvorkommen charakterisieren und klassifizieren lassen.

Viele der hier vorgestellten Konzepte beruhen auf einer strahlenseismischen Beschreibung, die eng an die geometrische Optik angelehnt ist und mit deren Hilfe sich eine amplitudenbewahrende Migration erst darstellen lässt. Allerdings muss berücksichtigt werden, dass die Strahlenseismik zur Beschreibung der Wellenausbreitung nur gültig ist, wenn die Wellen hinreichend hochfrequent sind. Dabei ist dieser Begriff nicht absolut zu verstehen, sondern es muss immer das Verhältnis zwischen Wellenlänge und den charakteristischen Größen des Untersuchungsobjektes betrachtet werden. In der strahlenseismischen Näherung lässt sich ein Reflexionsereignis in einem registrierten Seismogramm durch Gleichung (1.1) darstellen, wobei R_c der winkelabhängige Reflexionskoeffizient ist, $\mathcal{L}$ die sphärische Divergenz beschreibt und $\mathcal{A}$ alle anderen Faktoren repräsentiert, die auf die Amplitude einwirken können. Dazu zählen z. B. die Stärke der seismischen Quelle, Transmissionsverluste oder Dämpfung im Überbau des Reflektors, um nur einige zu nennen. Ferner stellt $F(t-\tau_R)$ das komplexe Quellsignal dar, das um die Reflexionslaufzeit τ_R verschoben ist. Eine komplexe Beschreibung wird gewählt, um Phasenverschiebungen während der Wellenausbreitung einfach berücksichtigen zu können. Mehrere Reflexionsereignisse in einem Seismogramm werden durch die Superposition einzelner Ereignisse – wie durch Gleichung (1.1) beschrieben – dargestellt.

Zur Beschreibung der räumlichen Position, an der ein Seismogramm aufgezeichnet wurde, wird der Vektor $\vec{\xi}$ verwendet, siehe Gleichung (1.2). Zusammen mit den Konfigurationsmatrizen $\mathbf{\Gamma}$, eingeführt in Gleichung (1.3), lässt sich so die Geometrie eines seismischen Experiments eindeutig beschreiben (Abbildung 1.6). Gängige Quell-Empfänger Konfigurationen sind dabei

- die *common-shot* (CS) Konfiguration. In einer CS Sektion sind Seismogramme enthalten, die an verschiedenen Empfängern mit zunehmendem Abstand zu einer einzigen seismischen Quelle aufgezeichnet wurden.

- die *common-offset* (CO) Konfiguration. Eine CO Sektion enthält alle Seismogramme, bei denen der Quell-Empfänger Abstand, der so genannte *offset*, konstant ist.

- die *zero-offset* (ZO) Konfiguration. Eine ZO Sektion ist eine spezielle CO Sektion, bei der für jedes Seismogramm die zugehörige Quell- und Empfängerposition zusammen fallen, der *offset* also Null ist.

- die *common-midpoint* (CMP) Konfiguration. Eine CMP Sektion enthält alle Seismogramme, die zwar mit unterschiedlichen *offset* aufgezeichnet wurden, aber immer den gleichen Mittelpunkt zwischen Quelle und Empfänger haben.

Abbildung 1.5 fasst die einzelnen Quell-Empfänger Konfigurationen graphisch zusammen.

[1] Seismische Quellen können dabei in der Regel als Punktquellen angenommen werden. Lokal betrachtet kann es während der Wellenausbreitung auch zu Fokussierungseffekten kommen, sodass die Energiedichte wieder zunimmt.

Es gibt viele Methoden, um seismische Daten zu migrieren. Diese können grob unterteilt werden in Methoden, die auf a) einer Integrallösung (Kirchhoff-Migration), b) einer Lösung basierend auf Ableitungen (Finite-Differenzen Wellengleichungsmigration) oder c) einer Lösung im Spektralbereich (Frequenz-Wellenzahl-Migration) der Wellengleichung beruhen. Zudem muss man zwischen 2D und 3D Migrationen, Zeit- und Tiefenmigrationen, sowie *prestack* und *poststack* Migrationen unterscheiden. Während bei einer *poststack* Migration lediglich eine (simulierte) ZO Sektion transformiert wird, so gehen in eine *prestack* Migration alle CO Sektionen ein (siehe dazu auch Abbildung 1.7). Die *prestack* Migration ist daher wesentlich aufwändiger, allerdings dafür auch in der Lage, migrierte Abbilder zu liefern, die für weitere Analysen wie z. B. AVA von Bedeutung sind. Eine Zeitmigration unterscheidet sich von einer Tiefenmigration primär durch die Domäne, in der das Migrationsresultat dargestellt wird: einmal im Zeit- und einmal im Tiefenbereich. Allerdings gibt es weitere fundamentale Unterschiede, die sich vor allem in dem zur Migration verwendeten Geschwindigkeitsmodell und damit verbundener Annahmen ausdrücken. Auf die Zeitmigration wird in dieser Arbeit nicht weiter eingegangen.

Nach den einleitenden Worten in Kapitel 1 wird in Kapitel 2 auf die Herleitung der Wellengleichung eingegangen, die allen seismischen Abbildungsverfahren zugrunde liegt. Die Strahlenseismik, von der in dieser Arbeit ausgiebig Gebrauch gemacht wird, wird in Kapitel 3 vorgestellt. Das allgemeine Problem der Inversion wird mathematisch in Kapitel 4 behandelt, bevor in Kapitel 5 die amplitudenbewahrende Kirchhoff-Migration erklärt wird. In Kapitel 6 wird der Zusammenhang zwischen der mathematischen Beschreibung von Migrationsartefakten und den geometrischen Konzepten der Kirchhoff-Migration hergestellt. Zum ersten Mal ist es hier gelungen, den bisher nur rein mathematisch beschriebenen Artefakten eine anschauliche geometrische Erklärung hinzuzufügen. In Kapitel 7 werden weitere Aspekte der Kirchhoff-Migration behandelt, bevor diese in Kapitel 8 anhand eines synthetischen Datenbeispiels quantitativ näher untersucht werden. Ein Ablauf zur seismischen Datenverarbeitung, der auf der *common-reflection-surface* (CRS) Stapelung beruht, wird in Kapitel 9 vorgestellt und anhand eines Realdatensatzes getestet. Im letzten Kapitel fasse ich diese Dissertation kurz zusammen.

Kapitel 2: Die Wellengleichung

Für ein kleines Volumenelement innerhalb eines Festkörpers lässt sich das zweite Newtonsche Gesetz wie in Gleichung (2.1) darstellen. Dabei ist $\vec{u}(\vec{x},t)$ der Verschiebungsvektor, t steht für die Zeit und $\rho(\vec{x})$ beschreibt die räumlich variierende Dichte des Körpers. Die Dichte der externen Kräfte, die auf das gesamte Volumen V des Körpers wirken, wird durch $\vec{f}$ beschrieben. Die Zugkraft, die über die Oberfläche S auf das Volumen V des Körpers wirkt, ist durch $\vec{T}(\vec{n})$ gegeben. Dabei ist $\vec{n}$ ein nach außen zeigender Normalenvektor der Randfläche. Durch den Satz von Gauß lässt sich die Integralgleichung (2.1) in die Bewegungsgleichung (2.2) umwandeln. Diese Gleichung stellt eine der grundlegendsten Gleichungen der gesamten Seismik und Seismologie dar, da sie die Kräfte in einem Medium mit messbaren Partikelverschiebungen in Verbindung bringt. Hierbei tritt der Cauchysche Spannungstensor $\underline{\tau}$ auf, der wiederum über das verallgemeinerte Hooksche Gesetz (2.3) mit dem Dehnungstensor $\underline{e}$ zusammenhängt. Der Tensor $\underline{\underline{c}}$ wird Elastizitätstensor genannt und beschreibt die Materialeigenschaften des Körpers. Bringt man den Dehnungstensor mit dem Verschiebungsvektor in Verbindung, so lässt sich aus der Bewegungsgleichung die elastodynamische Wellengleichung (2.7) für inhomogene, anisotrope, linear-elastische Medien gewinnen. In allgemeiner Form lässt sich diese Gleichung nicht lösen. Glücklicherweise sind die elastischen Eigenschaften der Erde oft nicht von

Richtung oder Orientierung abhängig, sodass sich die 21 unabhängigen Komponenten des Elastizitätstensors auf zwei reduzieren lassen: die Lamé Parameter. In diesem so genannten isotropen Fall vereinfacht sich die allgemeine elastodynamische Gleichung und kann wie in Gleichung (2.10) angegeben werden (hier in Vektornotation). Für den homogenen Fall, d. h. falls die Lamé Parameter unabhängig vom Ort durch eine konstante Größe gegeben sind, erhält man Gleichung (2.11). In einem homogenen, isotropen Medium gibt es zwei Wellentypen, die sich unabhängig voneinander ausbreiten: Das ist einerseits die longitudinal polarisierte P-Welle (Kompressionswelle), andererseits die S-Welle, die transversal polarisiert ist (Scherwelle). Beide Wellentypen koppeln nur an Schichtgrenzen und sind ansonsten völlig unabhängig voneinander. Obwohl die Erde sicherlich nicht homogen ist, kann man in nahezu allen seismischen Experimenten unabhängige P- und S-Wellenausbreitung beobachten. Das legt den Schluss nahe, dass man auch in schwach inhomogenen Medien P- und S-Wellen als nahezu unabhängig voneinander betrachten kann, zumindest solange die elastischen Eigenschaften des Mediums innerhalb einer Wellenlänge als konstant angenommen werden können. In der Explorationsgeophysik spielen auch die akustischen Wellengleichungen (2.25) bzw. (2.26) eine bedeutende Rolle. Sie beschreiben die Wellenausbreitung in Fluiden, die bei der Suche nach fossilen Brennstoffen oft als Näherung für die feste Erde verwendet werden.

Kapitel 3: Die Strahlenseismik

Die Strahlenseismik ist eine effiziente Methode, um Laufzeiten und Amplituden ausgewählter Wellentypen in inhomogenen Medien berechnen zu können. Sie basiert auf einer Hochfrequenzapproximation der Wellengleichung. Hochfrequent in diesem Zusammenhang bedeutet, dass die Wellenlänge des seismischen Signals klein gegenüber den charakteristischen Größen des Mediums sein muss. Mit Hilfe des Reihenansatzes (3.2), der so genannten *ray series*, lassen sich die Eikonalgleichung (3.4) und die Transportgleichung (3.5) herleiten. Sie stellen die zentralen Gleichungen der Strahlentheorie dar. Sehr oft beschränkt man sich auf die Terme nullter Ordnung der Reihenentwicklung (*zero-order ray theory*), da sie in den meisten Fällen bereits ausreichend genau die Wellenausbreitungsphänomene beschreiben und im Vergleich zu Termen höherer Ordnung relativ einfach handzuhaben sind.

Die Eikonalgleichung (3.4) lässt sich mit Hilfe der Methode der Charakteristiken lösen. Wendet man diese Methode an, so ergibt sich das (kinematische) *ray-tracing* System (3.7). Durch dieses Gleichungssystem lassen sich die Charakteristiken des Wellenfeldes, die so genannten Strahlen, finden. Die Laufzeiten ergeben sich durch einfache Integration von $ds/v(\vec{x})$ entlang der zuvor berechneten Strahlwege. Die Berechnung der Amplituden als Lösung der Transportgleichung gestaltet sich etwas aufwändiger und kann erst nach dem Lösen der Eikonalgleichung geschehen. Im Strahlkoordinatensystem zeigt sich schnell, dass die Amplitude (3.13) hauptsächlich durch den Parameter J (*ray jacobian*) bestimmt ist, der in einer geometrischen Deutung die Strahlendichte repräsentiert. Er ist eng verknüpft mit der sphärischen Divergenz bzw. dem *geometrical-spreading* Faktor $\tilde{\mathcal{L}}$.

Einen anderen Weg, geometrische Eigenschaften und damit auch die sphärische Divergenz zu bestimmen, bietet das paraxiale *ray-tracing* System (3.15), hier in strahlzentrierten Koordinaten angegeben. Eng damit verwandt ist das dynamische *ray-tracing* System. Von fundamentaler Bedeutung ist in diesem Zusammenhang die so genannte Propagatormatrix $\mathbf{\Pi}$, mit deren Hilfe sich alle relevanten Parameter bei der Wellenausbreitung bestimmen lassen. So lässt sich der *geometrical-spreading* Faktor $\tilde{\mathcal{L}}$ wie in Gleichung (3.23) durch die Untermatrix $\mathbf{Q}_2$ der Propagatormatrix ausdrücken. Der normalisierte *geometrical-spreading* Faktor $\mathcal{L}$ ist in Gleichung (3.24) angegeben. Er spielt, wie bereits angedeutet, eine wichtige Rolle bei der amplitudenbewahrenden Migration.

Kapitel 4: Seismische Inversion

Ein Anfangswert- und Randwertproblem für die akustische Wellengleichung lässt sich wie in Gleichung (4.2) formulieren, wobei der Vektor $\vec{x}$ in einem abgeschlossenen Raum Ω liegt und die Zeit zwischen t_0 und T variiert. Neben den Anfangsbedingungen zur Zeit $t = t_0$ muss auch eine Randbedingung auf der Fläche $\partial\Omega$, die das Volumen Ω abgrenzt, gegeben sein. Dies ist entweder die Dirichletsche Randbedingung, falls das Wellenfeld selbst auf $\partial\Omega$ angegeben wird, oder die von Neumann Randbedingung, falls die Ableitung des Wellenfeldes in Richtung der Flächennormalen gegeben ist. Eine Kombination aus beiden Randbedingungen, die ebenfalls möglich wäre, wird hier nicht betrachtet. Zur Berechnung der Greenschen Funktion wird im Quellterm in Gleichung (4.2) die Kraftdichte $f(\vec{x}, t)$ durch zwei Deltafunktionen $\delta(t-t_0)\delta(\vec{x}-\vec{x}_0)$ ersetzt, womit sich Gleichung (4.5) ergibt. Zudem müssen wiederum Anfangs- und Randbedingungen festgelegt werden, die im Falle der Greenschen Funktion durch homogene Gleichungen gegeben sind. So beschreibt die Greensche Funktion letztendlich die Entwicklung des Wellenfeldes mit fortlaufender Zeit, wenn zur Zeit t_0 eine Quelle am Ort $\vec{x}_0$ ausgelöst wird. Man sollte dabei in Erinnerung behalten, dass der Terminus „Quelle“ nicht unbedingt eine physikalisch vorhandene, seismische Quelle beschreibt. Ich komme später darauf zurück.

Die Frage stellt sich, ob das Anfangs- und Randwertproblem (4.2) stets von Neuem gelöst werden muss, falls sich die Anfangsbedingungen oder auch der Quellterm ändern. Dies ist zum Glück nicht der Fall, wie sich mit Hilfe der Greenschen Funktion und der Greenschen Formel zeigen lässt. Multipliziert man (von links) Gleichung (4.1) mit G und Gleichung (4.5) mit P und bildet die Differenz, so erhält man Gleichung (4.10). Diese über den Raum Ω und die Zeit t integriert unter Beachtung der Anfangsbedingungen für P und G liefert nach einigen Vereinfachungen das so genannte Kirchhoff-Theorem (4.15). Betrachtet man dieses Theorem im Hinblick auf ein seismisches Experiment, bei dem es keine Quellen im Gebiet Ω gibt und vor dem Auslösen der seismischen Quelle keine Wellen propagieren, so vereinfacht es sich zu Gleichung (4.17). Allerdings kann man bei einem seismischen Experiment nicht davon ausgehen, dass das Gebiet Ω einen geschlossenen und umrandeten Raum darstellt, schließlich ist oft nur ein Teil S der Randfläche $\partial\Omega$ zugänglich, siehe Abbildung 4.1. Stellt man sicher, dass von dem Teil S_* der Fläche $\partial\Omega$, der nicht zugänglich ist, keine Beiträge zur Integration kommen, so ergibt sich mit einer Greensche Funktion, welche die Dirichletsche Randbedingung erfüllt, Gleichung (4.18). Diese Gleichung ist als Kirchhoff-Integral bekannt. Für ein homogenes Medium lässt sich die Greensche Funktion analytisch berechnen und ist mit der Dirichletschen Randbedingung durch Gleichung (4.20) gegeben. Durch Einsetzen in das Kirchhoff-Integral erhält man nach einigen Vereinfachungen schließlich Gleichung (4.25). Eine weitere Vereinfachung ergibt sich durch die so genannte Fernfeld-Approximation, d. h. es wird nur das Wellenfeld in einer gewissen Entfernung von der seismischen Messoberfläche betrachtet. Hier wird das zweite Integral in Gleichung (4.25) aufgrund der höheren Ordnung in $1/r_0$ vernachlässigt.

Für das inverse Problem lässt sich ähnlich vorgehen wie für das direkte Problem, das bereits beschrieben wurde. Nach einigen Vereinfachungen erhält man Gleichung (4.31), die als Porter-Bojarski-Integral bekannt ist und den Kirchhoff-Migrationsprozess beschreibt. Wiederum kann für ein homogenes Medium durch Einsetzen der analytisch berechneten Greenschen Funktion eine weitere Vereinfachung erzielt werden, die auf Gleichung (4.36) führt. Diese Gleichung ist identisch mit der bekannten 3D Migrationsformel für gestapelte Sektionen von Schneider (1978).

Unabhängig davon, ob das direkte oder das inverse Problem betrachtet wird, kann mit Hilfe der eingeführten Integrale das Wellenfeld an einem beliebigen Ort $\vec{x}_0$ in Ω berechnet werden, wenn das Wellenfeld über einen gewissen Zeitraum auf einer geeigneten Fläche S bekannt ist. Dies wird erreicht, indem alle Beiträge von virtuellen Huygensschen Quellen, die in der Fläche S liegen, am Ort $\vec{x}_0$ in

Ω betrachtet werden. Dabei dient die Greensche Funktion als Verbindung zwischen dem Beobachtungspunkt und der Position der virtuellen Huygensschen Quelle. Mit Hilfe der Greenschen Funktion werden also alle Beiträge zum Wellenfeld am Ort $\vec{x}_0$ richtig gewichtet und abhängig von der Ausbreitungsgeschwindigkeit der Wellen zeitlich verzögert aufsummiert. Wie man hier sieht und wie bereits früher angedeutet wurde, muss der Terminus „Quelle" nicht einer physikalischen Quelle im Feld entsprechen. Damit man letztendlich ein migriertes Abbild erhält und nicht nur ein zurückpropagiertes Wellenfeld, muss zusätzlich eine Abbildungsbedingung (*imaging condition*) angewendet werden. Da mit Hilfe des Kirchhoff-Integrals keine Rückwärtspropagation von Wellen beschrieben werden kann, wurde als Gegenstück die Kirchhoff-Migration eingeführt, die eine Vorwärtspropagation des Wellenfeldes zurück in der Zeit realisiert. Das direkte und indirekte Problem unterscheidet sich daher hauptsächlich durch die Greensche Funktion. Für praktische Anwendungen bei realistischen Erdmodellen muss diese dabei stets nummerisch berechnet werden.

Kapitel 5: Amplitudenbewahrende Kirchhoff-Migration

Nach der strikt mathematischen Betrachtungsweise in Kapitel 4 wird nun eine anschaulichere Art der Kirchhoff-Migration vorgestellt, die auf geometrischen Betrachtungen beruht. Dabei spielen die Begriffe der Huygensfläche und Isochrone eine entscheidende Rolle. Die Huygensfläche τ_D, auch Diffraktionslaufzeitfläche genannt, ist das kinematische Abbild im Zeitbereich eines Diffraktionspunktes im Tiefenbereich, während die Isochrone, auch Fläche gleicher Reflexionslaufzeit genannt, das kinematische Abbild im Tiefenbereich eines Punktes im Zeitbereich ist. Beide Flächen lassen sich über die Summe von Laufzeiten entlang von Strahlästen von der Oberfläche zu Untergundpunkten M (und zurück) in einem gegebenen Geschwindigkeitsmodell und für eine bestimmte Quell-Empfänger Konfiguration konstruieren. Dabei gelten folgende Eigenschaften: a) Die zu einem Reflektorpunkt M_R zugehörige Huygensfläche τ_D und die Reflexionslaufzeitfläche τ_R sind tangential im Zeitbereich; b) die zu einem Reflexionsereignis gehörende Isochrone und der Reflektor sind tangential im Tiefenbereich, siehe auch Abbildung 5.1. Diese beiden Eigenschaften können als Hagedoorns Abbildungsbedingungen (Hagedoorn, 1954) aufgefasst werden. Die Bedingung, dass Flächen tangential sind, wird später mathematisch mit Hilfe der Methode der stationären Phase ausgewertet.

Die Kirchhoff-Tiefenmigration lässt sich mathematisch durch das Doppelintegral (5.2) darstellen. Geometrisch interpretiert bedeutet diese Gleichung eine Summation von Amplitudenwerten der (nach der Zeit abgeleiteten) Seismogramme entlang von Huygensflächen. Die so genannte Apertur A begrenzt dabei die Summationsfläche. Der auf diese Art erhaltene Summationswert wird dem Tiefenpunkt (Diffraktionspunkt) M, welcher der jeweiligen Huygensfläche zugeordnet ist, zugewiesen. Nur jene Tiefenpunkte, deren Huygensfläche tangential an die tatsächliche Reflexionslaufzeitfläche ist, werden Werte aufweisen, die nennenswert von Null verschieden sind. Letztendlich ergibt sich so ein Abbild des Untergrundes, das die Reflektorpositionen so korrekt wie es das verwendete Geschwindigkeitsmodell zulässt wiedergibt. Bei einer amplitudenbewahrenden Migration soll das Migrationsresultat aber nicht nur kinematisch korrekt sein, sondern die Amplituden im migrierten Abbild sollen ein Maß für den winkelabhängigen Reflexionskoeffizienten werden. Dies wird dadurch erreicht, dass der Effekt der sphärischen Divergenz durch einen Gewichtungsfaktor während der Summation rückgängig gemacht wird. Die Gewichtsfunktion kann mit Hilfe der Methode der stationären Phase im Frequenzbereich berechnet werden. Sie ist im allgemeinen Fall durch Gleichung (5.13) gegeben und scheint in dieser Notation von Reflektoreigenschaften abhängig zu sein. Durch geeignete Umformungen zu Gleichung (5.28) lässt sich aber zeigen, dass dies nicht der Fall ist. Erstaunlicherweise korrigiert die Summation bei der Kirchhoff-Migration bereits implizit den Teil des *geometrical-spreading* Faktors, der von den Reflektoreigenschaften abhängt, und die Gewichtsfunktion selbst entfernt lediglich die restlichen Effekte im Überbau des Reflektors. Für spezielle Messkonfigurationen vereinfacht sich die Gewichtsfunktion. Für die *zero-offset* Konfiguration ist sie besonders einfach und in Gleichung (5.30) dargestellt.

Ein besonderes Szenario stellt der so genannte 2.5D Fall dar. Hier betrachtet man die dreidimensionale Wellenausbreitung in einem Medium, dessen Eigenschaften nur zweidimensional variieren, d. h. entlang der ξ_2-Achse senkrecht zur seismischen Linie konstant sind. In so einem Fall lässt sich die Migrationsformel (5.2) vereinfachen, denn das Integral über ξ_2 kann aufgrund der Symmetrie des Modells analytisch ausgewertet werden. Man erhält schließlich Gleichung (5.20), die eine Kirchhoff-Migration im 2.5D Fall beschreibt. Man sieht leicht, dass hier nur noch entlang von Kurven und nicht mehr entlang von Flächen summiert werden muss, was graphisch in Abbildung 5.3 verdeutlicht ist. Dadurch ist eine derartige Migration bedeutend schneller, sie liefert allerdings auch nur dann korrekte Ergebnisse, wenn sich die Eigenschaften des Untergrundes wirklich nur zweidimensional (nämlich in Richtung der seismischen Linie) ändern. Eine Gewichtsfunktion für eine amplitudenbewahrende 2.5D Migration kann auf gleiche Art und Weise wie im 3D Fall bestimmt werden.

Eine Alternative zur Diffraktionsstapelung bietet die Kirchhoff-Migration, bei der Isochronen verwendet werden. Hier wird die Amplitude, die man an einem Punkt eines Seismogramms findet, entlang der zugehörenden Isochrone gewichtet verteilt – man spricht auch von verschmieren. Führt man dieses Verfahren für alle Punkte im Zeitbereich aus und addiert im Tiefenbereich stets alle Beiträge zu einem Tiefenpunkt auf, wenn eine Isochrone durch diesen Punkt verläuft, so ergibt sich wiederum durch konstruktive und destruktive Interferenz der Beiträge ein migriertes Abbild. Diese Methode der Migration ist zur Diffraktionsstapelung äquivalent und beruht auf dem Prinzip der Dualität, das von Tygel et al. (1995) gezeigt wurde.

Kapitel 6: Apertureffekte

Seismische Abbilder nach einer Kirchhoff-Migration sind oft von so genannten Migrationsrandeffekten begleitet, wenn die Implementierung des Algorithmus nicht sorgfältig vorgenommen wurde. Aufgrund ihres Erscheinungsbildes werden diese Effekte auch als *smile* bezeichnet. Sie entstehen dadurch, dass die Apertur (d. h. die Ausdehnung der Fläche, über die in der Kirchhoff-Migration summiert wird) begrenzt ist. Eine natürliche, feste Grenze stellt das Gebiet dar, in dem überhaupt Daten vorliegen, d. h. die Größe des Messgebietes ist eine Aperturbegrenzung. Oft ist es jedoch vorteilhaft, die Apertur bei der Summation entlang der Diffraktionslaufzeitflächen weiter einzuschränken, denn

1. je kleiner die Apertur, desto schneller der Algorithmus.
2. ein eingeschränkter Summationsoperator hilft, so genanntes *operator aliasing* zu verhindern.
3. je weniger irrelevante Daten entfernt von Nutzsignalen aufsummiert werden, desto weniger Rauschen tritt im migrierten Abbild auf.

Die beste Methode, eine Apertur einzuschränken, ist modellbasiert und nutzt die projizierte Fresnelzone, das ist die Projektion der Fresnelzone am Reflektor in der Tiefe entlang von Strahlen hin zur Messoberfläche. Leider ist die Bestimmung dieser projizierten Fresnelzone vor der Migration nur schwer möglich, weswegen üblicherweise ein Kompromiss eingegangen und eine feste (tiefenabhängige) Apertur bei der Kirchhoff-Migration vorgegeben wird. Diese Einschränkung der Apertur verursacht bei unzureichender Implementierung die genannten Artefakte im Migrationsresultat, die bisher in der Literatur nur rein mathematisch beschrieben wurden. In diesem Kapitel wird eine Beziehung zwischen den Ergebnissen der Methode der stationären Phase, mit deren Hilfe sich die Randeffekte mathematisch beschreiben lassen, und einfachen geometrische Situationen hergestellt. So ergibt sich ein intuitives Verständnis für die Apertureffekte und die Methoden, diese zu verhindern.

Die bereits erwähnte Integralformel für eine 2.5D Migration lässt sich im Frequenzbereich durch Gleichung (6.1) darstellen. Auch dieses Integral kann nicht analytisch gelöst werden. Allerdings kann mit Hilfe der Methode der stationären Phase eine Näherung für das Resultat angegeben werden. Die genannte Methode ist in Anhang A ausführlich beschrieben. Nach deren Anwendung erhält man die Approximation des Migrationsresultats (6.4) und nach der Rücktransformation in den Zeitbereich Gleichung (6.7). Hier zeigt sich, dass der Wert, der einem Tiefenpunkt in der Kirchhoff-Migration zugewiesen wird, aus drei Hauptanteilen besteht: Einem Anteil, der vom so genannten stationären Punkt stammt und in der Regel dominiert. Der stationäre Punkt ist bereits aus Kapitel 5 bekannt und entspricht dort der Stelle, an der die Diffraktions- und die Reflexionslaufzeitfläche (bzw. -kurve im Falle einer 2.5D Migration) identische erste Ableitungen aufweisen. Der Anteil, der vom stationären Punkt stammt, stellt das eigentliche migrierte Reflektorabbild dar. Hinzu kommen zwei weitere Anteile, die vom linken und rechten Rand der Apertur stammen. Sie verursachen die Migrationsrandeffekte. Im Folgenden werden diese mathematischen Terme mit Hilfe von Abbildung 6.2 zu einfachen geometrischen Situationen in Verbindung gesetzt.

Der eigentliche Reflektor wird durch Punkte wie M_1 aufgebaut. Die zugehörenden Huygenskurven sind stets tangential zur Reflexionslaufzeitkurve; es kommt daher zu konstruktiver Interferenz beim Aufsummieren und der Wert, der M_1 zugewiesen wird, ist durch den Hauptterm in Gleichung (6.7) gegeben. Die Endpunkte des Summationsoperators liegen in einem Gebiet, in dem die Seismogramme eine Amplitude Null haben, d. h. die Randterme in Gleichung (6.7) sind Null.

Punkte wie M_2 unterscheiden sich kaum von Punkten wie M_1. Solange der Apex der Diffraktionslaufzeitkurve von M_2 innerhalb des Signals der Reflexionslaufzeit im Zeitbereich liegt, kommt es zu konstruktiver Interferenz; auf diese Art und Weise rekonstruieren Punkte wie M_2 letztendlich das Signal im Tiefenbereich und bauen das Reflektorabbild auf.

Der Punkt M_3 bildet den Rand des Reflektorabbildes im Tiefenbereich. Im Prinzip ist der Punkt M_3 äquivalent zu Punkt M_1, jedoch liegt nur der halbe Summationsoperator innerhalb der Daten im Zeitbereich und der stationäre Punkt (welcher der Position des Apex des Operators entspricht) liegt direkt am Rand. Rein geometrisch ergibt sich, dass einem Punkt M_3 genau der halbe Wert zugewiesen wird, den ein Punkt M_1 erhält, und diese geometrische Beobachtung wird durch die Methode der stationären Phase bestätigt.

Punkte wie M_4 haben quasi eine Amplitude Null im Tiefenbereich. Die zugehörenden Huygenskurven gehen durch das Reflexionssignal im Zeitbereich komplett hindurch und beim Aufsummieren kommt es daher zu destruktiver Interferenz. Mathematisch betrachtet liefern alle Terme in Gleichung (6.7) Null, denn sowohl der Apex der Summationskurve am Punkt stationärer Phase als auch die Endpunkte des Operators liegen außerhalb des Signals im Zeitbereich.

Bei einem Punkt wie M_5 liegen die Endpunkte des Summationsoperators im Signal der Reflexionsantwort im Zeitbereich. Es fehlen daher beim Aufsummieren entlang der Kurve Informationen für eine komplette destruktive Interferenz und als Konsequenz wird einem Punkt M_5 ein von Null abweichender Wert zugewiesen. Es bildet sich so ein Migrationsartefakt aus, was auch mathematisch durch die Terme in Gleichung (6.7) prognostiziert wird. Zwar ist der Hauptterm hier Null, da sich der Apex des Summationsoperators außerhalb des Signals im Zeitbereich befindet, aber beide Randterme liefern ein von Null abweichendes Ergebnis.

Für Punkte wie M_6 ist das Verhalten ähnlich, allerdings liegt hier nur ein Operatorendpunkt innerhalb des Signals. Aus geometrischen Gründen sollte deshalb die Amplitude an einem Punkt M_6 genau der Hälfte der Amplitude bei M_5 entsprechen, was wiederum durch die Auswertung mit Hilfe der Methode der stationären Phase bestätigt wird. Der Punkt M_7 stellt den Übergang zwischen beiden Situationen dar.

Der wohl bekannteste Migrationsrandeffekt, der *smile*, wird durch Punkte wie M_8 und M_9 aufgebaut. Die zugehörenden Summationsoperatoren schneiden das Reflexionssignal im Zeitbereich genau am Rand der Daten, sodass es wiederum zu einer unvollständigen destruktiven Interferenz kommt und ein von Null abweichender Wert als Migrationsresultat bestehen bleibt. Die räumliche Position dieses Artefaktes entspricht dabei genau der Isochrone von Punkt P, siehe die Abbildungen 6.2 und 6.3(b). Das Vorzeichen der Amplituden entlang der beiden Teiläste des *smile* ist invers, sodass sich als Summe der Amplituden von zwei sich gegenüber liegenden Punkten M_8 und M_9 stets etwa Null ergibt, was in Abbildung 6.3(b) graphisch dargestellt ist. Mathematisch wird der Randeffekt wiederum durch einen der beiden letzten Terme in Gleichung (6.7) beschrieben. Die unterschiedlichen Vorzeichen der Amplituden beider Teiläste ergeben sich durch die Ableitung in den Randtermen von Gleichung (6.7), d. h. die Ableitung ist einmal positiv und einmal negativ, wie man sich anhand der Operatorsteigung am Punkt P in Abbildung 6.2 leicht verdeutlichen kann. Das frequenzabhängige Amplitudenverhalten, das die Auswertung mittels der Methode der stationären Phase vorhersagt, kann in diesem Beispiel ebenfalls nummerisch bestätigt werden, siehe Abbildung 6.4.

Alle hier genannten Artefakte treten natürlich nicht nur bei einer *poststack*, sondern auch bei einer *prestack* Migration auf. Ein Beispiel hierfür zeigt Abbildung 6.5. Die Frage stellt sich, wie diese Randeffekte, die ein Migrationsabbild nachhaltig beeinträchtigen können, bei einer Kirchhoff-Migration zu vermeiden sind. Mathematisch gesehen müssen die Randterme in Gleichung (6.7) verschwinden. Das wird durch so genanntes *tapern* erreicht: Weder der Summationsoperator noch die Daten dürfen am jeweiligen Ende abrupt abbrechen, stattdessen sollen sie langsam und stetig auslaufen. Ist der Bereich, in dem die *taper*-Funktion angewendet wird, richtig berechnet, so verschwinden die Migrationsartefakte, wie es in Abbildung 6.8 anhand eines Datenbeispiels beobachtet werden kann. Gleichung (6.10) gibt dabei für einfache Situationen die Länge des *taper*-Bereiches an. Auf die hier genannte Art und Weise lassen sich so sowohl die Apertureffekte, die durch einen begrenzten Operator, als auch die Effekte, die durch eine begrenzte Datenakquisition entstehen, in der Kirchhoff-Migration unterdrücken.

Kapitel 7: Weitere Aspekte der (Kirchhoff-)Migration

Die Kirchhoff-Tiefenmigration ist eine der ältesten Methoden, Daten vom Zeit- in den Tiefenbereich zu transformieren. Obwohl es heutzutage andere Methoden gibt, die in manchen Situationen qualitativ bessere Abbilder liefern können, wird die Kirchhoff-Migration immer noch sehr häufig verwendet. Einige Gründe hierfür sind:

- Die Kirchhoff-Migration ist sehr anschaulich und kann geometrisch leicht verstanden werden.
- Es existiert eine fundierte mathematische Beschreibung der Kirchhoff-Migration, wie sie auch in Kapitel 4 präsentiert wurde.
- Kirchhoff-Migration kann für beliebige Geschwindigkeitsmodelle angewendet werden und liefert zuverlässige Resultate, bleibt dabei aber recht effizient. Das bedeutet, Kirchhoff-Migration ist oft schneller als andere Migrationsmethoden, was ein wichtiger Faktor bei der industriellen Anwendung ist.

- Die Methode kann zielorientiert angewendet werden, d. h. es ist möglich, nur einen kleinen Bereich des Untergrundes direkt abzubilden, was einen weiteren Zeitvorteil gegenüber anderen Migrationsmethoden schaffen kann.

- Die Kirchhoff-Migration kann unmittelbar Daten handhaben, die mit einer irregulären Messgeometrie entlang von Flächen mit topographischen Änderungen aufgezeichnet wurden.

Natürlich hat die (amplitudenbewahrende) Kirchhoff-Migration auch Nachteile:

- Viele Implementierungen der Kirchhoff-Migration beruhen auf Strahlenseismik und unterliegen damit allen Einschränkungen, die diese Hochfrequenzapproximation mit sich bringt. Ferner wird oft nur ein möglicher Strahlweg, einen Untergrundpunkt mit einem Punkt an der Messoberfläche zu verbinden, berücksichtigt. Je nach Geschwindigkeitsmodell kann es aber mehrere dieser Wege geben, die alle berücksichtigt werden müssten. Dieses so genannte *multipathing*-Problem kann durch eine aufwändigere Art der Implementierung der Kirchhoff-Migration umgangen werden.

- Die Kirchhoff-Migration versagt manchmal bei der Abbildung komplexer Strukturen im Untergrund. Dieses Problem ist eng mit den zur Migration benötigten Laufzeittabellen und deren Erstellung verknüpft. Es gibt viele Methoden, Laufzeittabellen zu erzeugen, die sich hinsichtlich Schnelligkeit und Genauigkeit unterscheiden.

- Es tritt das so genannte *operator aliasing* auf. In der Kirchhoff-Migration werden die (diskretisierten) Daten unabhängig von ihrem Frequenzgehalt aufsummiert. Es kann dabei passieren, dass die steileren Teile der Huygensfläche das zu summierende Signal nicht mehr korrekt über benachbarte Seismogramme erfassen, d. h. es kommt zu Phasensprüngen der Signale beim Aufsummieren, was unter dem Begriff *aliasing* bekannt ist. Durch geeignete Frequenzfilter kann dieses Problem weitgehend umgangen werden, allerdings nimmt man dabei eine Verschlechterung der Auflösung im migrierten Abbild in Kauf.

- Die Berechnung von Gewichtsfunktionen für eine amplitudenbewahrende Migration ist aufwändig. Allerdings existieren Ansätze, die exakte Gewichtsfunktion durch einfache Ausdrücke zu approximieren und so den Aufwand zur Berechnung zu vermindern. Wie von Vanelle and Gajewski (2002a,b) gezeigt wurde, lassen sich die Gewichtsfunktionen auch vollständig aus Laufzeiten berechnen.

Auf einige Punkte wird im Folgenden noch näher eingegangen.

Die Berechnung der Greenschen Funktion ist ein wichtiger Teil einer Kirchhoff-Migration. Da die Greensche Funktion in zwei Teile zerlegt werden kann, nämlich einen Anteil, der mit Laufzeiten zu tun hat, und einen zweiten Anteil, der sich mit Amplituden befasst, können auch die Methoden zur Berechnung der Greenschen Funktion in zwei Klassen unterteilt werden: Methoden, die lediglich Laufzeiten berechnen, und Methoden, welche die komplette Greensche Funktion, d. h. Laufzeiten und Amplituden, berechnen. Die meisten Implementierungen beruhen auf kinematischem oder dynamischem *ray tracing*, jedoch kann z. B. auch eine direkte Integration der Eikonalgleichung erfolgen. Eine besonders effiziente Methode, die Greensche Funktion im Rahmen der Kirchhoff-Migration zu bestimmen, stellt die *wavefront construction method* (Vinje et al., 1993, 1996a,b) dar. Hierbei werden nicht einzelne Strahlen in einem Geschwindigkeitsmodell berechnet, sondern ganze Wellenfronten.

Da Laufzeittabellen sehr viel Speicherplatz benötigen, werden einige Vereinfachungen gemacht: So werden z. B. die Laufzeiten nur auf einem groben Raster berechnet, das dann während der Kirchhoff-Migration auf ein enges Raster interpoliert wird. Die Berechnung und die Speicherung der Laufzeittabellen oder auch zusätzlicher dynamischer Parameter für eine amplitudenbewahrende Migration soll schnell und effizient erfolgen – dieser Punkt stellt quasi den Schlüssel zum Erfolg beim Abbilden von seismischen 3D Daten in der industriellen Anwendung dar.

Wie bereits erwähnt, sind die Amplituden eines Reflektorabbildes nach einer amplitudenbewahrenden Migration ein Maß für den winkelabhängigen Reflexionskoeffizienten. Dies kann man sich zunutze machen, wenn nach einer *prestack* Migration alle Spuren zusammengefasst werden, die ein festgelegte laterale Position widerspiegeln, aber von unterschiedlichen migrierten *offset*-Sektionen stammen. Diese Spuren bilden ein so genanntes *common-image gather* (CIG). Werden die Amplituden entlang eines Ereignisses in einem CIG extrahiert, so erhält man eine AVO-Kurve. Die Abkürzung AVO steht dabei für *amplitude-variation-with-offset*. Durch Umrechnen des *offset* in den Reflexionswinkel am Reflektor, erhält man schließlich die gesuchte Kurve für den winkelabhängigen Reflexionskoeffizienten. Durch eine Vereinfachung, die auf einer Idee von Shuey (1985) basiert, lassen sich zwei charakteristische Größen bestimmen, *intercept* und *gradient*. Mit ihrer Hilfe lassen sich Eigenschaften des Untergrundes näher bestimmen, was insbesondere bei Lagerstätten eine entscheidende Rolle spielen kann. AVO bzw. AVA Analysen sind nicht immer möglich; gelingt es jedoch, entsprechende Amplituden aus migrierten Abbildern zu extrahieren, so können dadurch wichtige Informationen weit über das rein strukturelle Abbild hinaus gewonnen werden.

Die Bestimmung eines Geschwindigkeitsmodells für die Migration stellt neben der Berechnung der Greenschen Funktionen einen weiteren wichtigen Punkt dar. Das Geschwindigkeitsmodell kann relativ glatt sein, es sollte jedoch die Laufzeiten im Untergrund so genau wie möglich wiedergeben. Es gibt eine Vielzahl von Geschwindigkeitsanalyseverfahren, mit denen ein erstes Modell zur Migration erstellt werden kann. Dieses Modell wird üblicherweise nach einer *prestack* Migration verfeinert und geht dann erneut in die Migration ein. Dieses iterative Verfahren wird dadurch ermöglicht, dass es nach einer *prestack* Migration eine Möglichkeit zur Evaluierung des Geschwindigkeitsmodells gibt. Das Kriterium für ein korrektes Geschwindigkeitsfeld ist dabei recht einfach: die Reflektoren im migrierten Abbild müssen in allen einzeln migrierten CO Sektionen stets in den gleichen Tiefen liegen. Ist das nicht der Fall, so sind Ereignisse in einem CIG nicht flach und das Geschwindigkeitsmodell, das für die Migration verwendet wurde, muss korrigiert werden. Hierfür gibt es einige Methoden, welche die Krümmung der Ereignisse in CIG auswerten und in eine Verfeinerung des ursprünglichen Geschwindigkeitsmodells umsetzen.

Das Auftreten von *aliasing* bei der Kirchhoff-Migration wurde bereits angesprochen. Im Prinzip muss man zwischen drei Arten von *aliasing* unterscheiden:

1. *aliasing* im Datenbereich entsteht, wenn die räumliche Messgeometrie und die zeitliche Aufzeichnung der Daten nicht zu den Reflektorneigungen im Untergrund und dem Frequenzgehalt der Signale passen. Diese Art von *aliasing* kann quasi nur durch geeignete Datenakquisition verhindert werden, nicht aber bei der Datenverarbeitung.

2. *aliasing* im Bildbereich tritt auf, wenn das Gitter, auf dem durch Summation im Zeitbereich das Migrationsergebnis berechnet wird, zu grob ist, um Reflektorneigungen korrekt wiederzugeben. Diese Art von *aliasing* kann bei der Kirchhoff-Migration einfach durch ein ausreichend enges Gitter an Diffraktionspunkten, die später das Migrationsbild aufbauen, vermieden werden.

3. *operator aliasing* entsteht, wenn die Steigung des Summationsoperators nicht zum Frequenzgehalt und den räumlichen Abständen der Seismogramme passt.

Das letztgenannte Phänomen ist im Gegensatz zu den beiden anderen beschränkt auf die Kirchhoff-Migration und kann die Abbildungsqualität drastisch reduzieren, wenn bei der Implementierung keine Vorkehrungen zur Vermeidung getroffen werden. Eine Möglichkeit besteht darin, zwischen real gemessenen Seismogrammen künstlich Seismogramme zu interpolieren, denn dadurch verringert sich der Spurabstand und bei der Summation entlang des Operators werden Phasensprünge der Signale verhindert. Dieses Verfahren ist aber sehr aufwändig und bedarf in der Regel, dass Ereignisse im Zeitbereich klar zu identifizieren sind, d. h. das Signal-zu-Rauschen (S/N) Verhältnis entsprechend hoch ist. Üblicherweise wird daher eine andere Methode zur Vermeidung von *operator aliasing* gewählt, die auf *anti-aliasing* Filtern (AAF) beruht: Abhängig von der Steigung der Diffraktionslaufzeitfläche werden die Seismogramme derart gefiltert, dass Phasensprünge beim Aufsummieren nicht mehr auftreten können. Es gibt verschiedene Implementierungen dieser Technik, die unterschiedliche Vor- und Nachteile haben, aber alle recht ähnlich funktionieren. Es sollte jedoch angemerkt werden, dass sich eine amplitudenbewahrende Migration und AAF quasi ausschließen, da durch die Filteroperation die Amplituden in den migrierten Abbildern verfälscht werden, wie z. B. von Zhang et al. (2001) gezeigt wurde.

Die Verarbeitung von seismischen Daten, die mit einer irregulären Messgeometrie entlang der topographiebehafteten Erdoberfläche aufgezeichnet wurden, stellt für einige Verfahren ein großes Problem dar. Diese Verfahren erwarten oft, dass Daten auf einer ebenen Fläche aufgenommen wurden und dabei die Messgeometrie regulär war. Da das bei Feldexperimenten insbesondere zu Land quasi nie der Fall ist, gibt es verschiedene Methoden, z. B. statische Korrekturen oder *redatuming*, um derartige Daten zu simulieren. Es kann jedoch Vorteile haben oder sogar notwendig sein, die Daten direkt von der topographischen Oberfläche in den Tiefenbereich zu migrieren. Die Kirchhoff-Migration ist hierzu in der Lage, wenn bei der Berechnung der Laufzeiten und Gewichte die tatsächliche Topographie und Messgeometrie berücksichtigt werden. Spurabstände müssen dabei stets lokal für jede Position bestimmt und können nicht über einen größeren Bereich gemittelt werden, da das zu starken Amplitudenschwankungen im migrierten Abbild führen kann. Diese Problematik ist unter dem Namen *acquisition footprint* in der Literatur bekannt. Als geeignetes Maß zur Gewichtung der Spurabstände bei 3D Daten zeigt sich die so genannte Voronoi-Zelle, siehe Abbildung 7.4. Für eine Spurposition definiert sich diese Zelle dadurch, dass alle Punkte in ihrem Innern näher an der gerade betrachteten Spurposition liegen als an allen anderen. Die Größe der Zelle ist somit ein Maß für die Verteilung der Nachbarspuren in der Umgebung. Durch die Gewichtung mit der Fläche der Voronoi-Zelle werden Amplitudenfluktuationen im migrierten Abbild drastisch vermindert und Analysen wie AVO oder AVA verbessert oder gar erst ermöglicht.

Die Topographie der Messoberfläche kann durch eine geeignete Triangulation, die zur Berechnung der Voronoi-Zellen durchgeführt werden muss, beschrieben werden. Der Gradient der Messoberfläche, der zur Bestimmung der Gewichtsfunktion bei einer amplitudenbewahrenden Migration benötigt wird, kann somit durch geeignete Mittelung von Normalenvektoren angrenzender Dreiecke der Triangulation näherungsweise bestimmt werden. Die Berücksichtigung der Topographie und der irregulären Messgeometrie in einer amplitudenbewahrenden Kirchhoff-Migration stellt also die Qualität der Amplitudeninformation nach der Migration sicher.

Kapitel 8: Synthetisches Datenbeispiel

Als Teil dieser Dissertation wurde ein Programm zur amplitudenbewahrenden Kirchhoff-Tiefenmigration geschrieben. Alle bisher vorgestellten Aspekte, z. B. die Berechnung der Gewichtsfunktion, die Aperturbegrenzungen und die *taper*-Funktionen sowie die Handhabung von Topographie und irregulären Messgeometrien, wurden dabei berücksichtigt. In diesem Kapitel wird die Qualität der vorgestellten Techniken anhand eines synthetischen Datenbeispiels verifiziert.

Hierfür wurde das in Abbildung 8.1 gezeigte Modell erstellt und entlang der topographischen Messoberfläche ein synthetischer Datensatz durch *ray tracing* simuliert. Die Abbildungen 8.2 (rechts) und 8.3 zeigen, dass sowohl die *poststack* als auch die *prestack* Migration in der Lage war, alle Reflektoren im Untergrund korrekt zu positionieren. Bei der *prestack* Migration wurden alle CO Sektionen einzeln migriert und anschließend im Tiefenbereich aufsummiert. Da sich statistisch verteiltes Rauschen beim Aufsummieren von N Sektionen nur um den Faktor $\sqrt{N}$ erhöht, während die Amplituden der kohärenten Reflektorabbilder um den Faktor N ansteigen, ist das S/N Verhältnis im migrierten Abbild nach der *prestack* Migration höher als bei der *poststack* Migration, wo nur eine einzelne verrauschte ZO Sektion migriert wurde. Beim Simulieren von ZO Sektionen bei realen Datensätzen ist es daher für ein klares Abbild nach der *poststack* Migration nötig, ein möglichst hohes S/N Verhältnis zu erzielen. Ich komme darauf später noch einmal zurück. Zwei CIG sind in Abbildung 8.4 gezeigt; hier sieht man, dass das bei der Kirchhoff-Migration verwendete Geschwindigkeitsmodell korrekt und konsistent mit den Daten war, denn alle Ereignisse im CIG sind flach. Extrahiert man aus den CIG die Amplituden des obersten Ereignisses, welche zu einem Punkt des obersten Reflektors im Modell aus Abbildung 8.1 gehören, und rechnet den *offset* in den Reflexionswinkel für diesen Reflektorpunkt um, so erhält man die in Abbildung 8.5 gezeigten AVA Kurven. Die extrahierten Amplituden stimmen sehr gut mit den theoretisch erwarteten Werten überein, was impliziert, dass die Gewichtsfunktion bei der amplitudenbewahrenden Migration erfolgreich die Effekte der sphärischen Divergenz rückgängig machen konnte.

Die Effekte des *acquisition footprint* wurden anhand von simulierten Datensätzen mit irregulären Messgeometrien sowohl für eine 2.5D als auch eine 3D Tiefenmigration getestet. Wie in Abbildung 8.6 zu sehen ist, entspricht die Amplitude entlang des ersten Reflektors nach einer amplitudenbewahrenden 2.5D *poststack* Migration nicht nur genau der prognostizierten akustischen Impedanz, sondern durch die Berücksichtigung lokaler Spurabstände können auch Amplitudenfluktuationen weitgehend vermieden werden. Abbildung 8.8 zeigt die Resultate für eine amplitudenbewahrende 3D *poststack* Migration. Auch hier werden zuverlässige Amplitudeninformationen erreicht, wenn Spurabstände bei der Diffraktionsstapelung mit den Flächen der in Abbildung 8.7 gezeigten Voronoi-Zellen gewichtet werden.

Kapitel 9: Ein Arbeitsablauf zum Abbilden seismischer Reflexionsdaten basierend auf der CRS Stapelung

Die datenorientierte *common-reflection-surface* (CRS) Stapelung (Müller, 1999; Jäger et al., 2001; Mann, 2002) entspricht einem generalisierten, mehrdimensionalen und mehrparametrigen Geschwindigkeitsanalyseverfahren. In erster Linie wurde es entwickelt, um ZO Sektionen mit hohem S/N Verhältnis zu simulieren, wie sie für gute Migrationsresultate nach einer *poststack* Migration nötig sind. Dieser Punkt wurde bereits früher angesprochen. Die CRS Methode stellt somit eine Alternative zur konventionellen Vorgehensweise mit *normal-moveout* und *dip-moveout* Korrekturen und anschließender Stapelung dar. Die CRS Technologie bietet zahlreiche weitere Vorteile, denn neben der simulierten

ZO Sektion erhält der Anwender zudem einen Satz von so genannten kinematischen Wellenfeldattributen. Diese sind im weiteren Verlauf der Datenverarbeitung sehr hilfreich: Sie können dazu benutzt werden, ein (erstes) Geschwindigkeitsmodell für die Tiefenmigration zu ermitteln. Diesem Modell kommt, wie bereits in früheren Abschnitten erwähnt wurde, eine große Bedeutung zu; je näher das ermittelte Geschwindigkeitsmodell dem wahren Modell kommt, desto schneller und einfacher lässt sich ein gutes Resultat bei der *prestack* Migration erzielen. Ferner können mit Hilfe der Attribute die sphärische Divergenz oder die Größe der projizierten Fresnelzone abgeschätzt werden. Mit der CRS Stapelung lässt sich somit ein Arbeitsablauf definieren, der von den vorverarbeiteten Daten im Zeitbereich bis hin zu einem Abbild im Tiefenbereich und weiteren Analysen wie z. B. AVA reicht.

Der hier vorgestellte Arbeitsablauf wird insbesondere im Hinblick auf die Migration anhand eines Realdatenbeispiels verdeutlicht. Die Daten wurden in der Nähe von Karlsruhe entlang von zwei seismischen Linien mit jeweils ungefähr 12 km Länge gewonnen. Die Messungen wurden durch die Deutsche Montan Technologie GmbH, Essen im Auftrag von HotRock EWK Offenbach/Pfalz GmbH, Karlsruhe durchgeführt. Die seismische Quelle bildeten drei Vibratoren, einer davon ist in Abbildung 9.3 gezeigt. Sie generierten Signale von etwa 10 s Länge im Frequenzbereich zwischen 12 und 100 Hz. Das Experiment diente dazu, ein detailgenaues strukturelles Abbild bis in eine Tiefe von ca. 3000 m zu erhalten. Ferner sollen geplante AVO bzw. AVA Analysen (litho)stratigraphische Informationen über den Untergrund liefern.

Die CRS Stapelung simuliert wie bereits erwähnt eine ZO Sektion. Im Gegensatz zu konventionellen Verfahren passt diese Methode bei 2D Datenakquisition allerdings ganze Flächen und nicht nur Trajektorien an die Daten an. Daher tragen wesentlich mehr Seismogramme zur ZO Simulation bei, was zu einem im Vergleich zu den konventionellen Methoden höheren S/N Verhältnis führt. Der Stapeloperator in 2D ist für ein ZO *sample* (x_0, t_0) durch Gleichung (9.1) gegeben, wobei der halbe *offset* h und der Mittelpunkt zwischen Quelle und Empfänger durch x_m gegeben sind. Die verbliebenen Parameter α, R_{N} und R_{NIP} sind die so genannten kinematischen Wellenfeldattribute; sie entsprechen einem Auftauchwinkel und zwei Krümmungsradien von Wellenfronten. Die Attribute werden durch Kohärenzanalysen bestimmt, indem sämtliche realistische Kombinationen von Parametern durchprobiert werden. Der Operator, der sich letztendlich am Besten an das Reflexionsereignis im Zeitbereich anpasst, liefert die endgültigen Attribute, die zur Stapelung verwendet werden. Die CRS Stapelung wurde auf die oben erwähnten Realdaten angewendet. Die simulierten ZO Sektionen sind in Abbildung 9.5 für beide seismische Linien dargestellt. Die Abbildungen 9.6(a) - (c) zeigen für eine seismische Linie die optimalen kinematischen Wellenfeldattribute, die für die Stapelung mit dem Operator (9.1) verwendet wurden; zusätzlich ist in (d) die Kohärenz abgebildet.

Die Wellenfeldattribute wurden für eine tomographische Inversion (Duveneck, 2004) herangezogen. Dabei macht man sich zu Nutzen, dass die mit dem Parameter R_{NIP} assoziierte Welle in einem korrekten Geschwindigkeitsmodell zur Zeit $t = 0$ am Reflektor fokussieren muss. Nach dem Auswählen geeigneter Datenpunkte in der Kohärenzsektion und dem gleichzeitigen Extrahieren der zugehörenden Wellenfeldattribute wurde iterativ der Fehler zwischen diesen Attributen und vorwärts berechneten Attributen minimiert. Als Ergebnis der tomographischen Inversion erhielt ich die in Abbildung 9.7 gezeigten glatten Geschwindigkeitsmodelle. Diese wiederum wurden für die Berechnung der Laufzeittabellen in einer anschließenden Kirchhoff-Tiefenmigration verwendet.

Die Ergebnisse der *poststack* Kirchhoff-Tiefenmigration sind in Abbildung 9.8 und die entsprechenden Resultate einer *prestack* Migration in Abbildung 9.9 zu sehen. Bereits mit bloßem Auge lassen sich zahlreiche Verwerfungen erkennen, deren Anzahl alle Erwartungen übertrifft. Alte Daten aus einem heute bereits verfüllten Bohrloch in der Nähe einer der seismischen Linien zeigt eine gute

Übereinstimmung zu den Tiefenlagen von Schlüsselhorizonten im migrierten Abbild. Eine vorläufige strukturelle Interpretation der Ergebnisse ist in Abbildung 9.11 gezeigt. Für den *interpreter* (N. Harthill; pers. Komm.) ergeben sich anhand dieser Ergebnisse einige Vorteile gegenüber den Ergebnissen (hier nicht gezeigt) eines konventionelles Arbeitsablaufes, der ebenfalls durchgeführt wurde:

- Reflektoren werden generell besser abgebildet.
- Laterale Änderungen von Reflektoreigenschaften sind leicht zu erkennen.
- Reflektoren in 2.000 m Tiefe und darunter sind klar abgebildet.
- Klüfte und Verwerfungen sind in allen Tiefen klar abgebildet und können teilweise von der Oberfläche aus bis in Tiefen von nahezu 3.000 m verfolgt werden.

Für dieses Datenbeispiel hat der hier vorgestellte Arbeitsablauf, bestehend aus der CRS Stapelung, der Bestimmung eines Geschwindigkeitsmodells mittels einer attributbasierten tomographischen Inversion sowie *prestack* und *poststack* Kirchhoff-Tiefenmigrationen, sehr gute Ergebnisse geliefert. Zu betonen ist, dass die meisten Arbeitsabläufe weitgehend automatisiert sind und somit nur geringe Eingriffe durch den Anwender erforderlich sind.

Kapitel 10: Zusammenfassung

In dieser Dissertation wurde die Transformation von seismischen Reflexionsdaten vom Zeit- in den Tiefenbereich näher untersucht. Zunächst wurden die physikalischen und mathematischen Grundlagen zur Beschreibung der elastischen bzw. akustischen Wellenausbreitung präsentiert, auf die in späteren Kapiteln immer wieder zurück gegriffen wurde. Nach einer strikt mathematischen Herleitung der Inversion seismischer Daten bin ich auf die geometrische Beschreibung sowie die Berechnung von Gewichten für eine amplitudenbewahrende Migration eingegangen. Auf Grundlage der geometrischen Betrachtungen ist es mir gelungen, für die bisher in der Literatur nur rein analytisch behandelten Apertureffekte eine anschauliche Erklärung zu finden. Somit wurde eine Lücke geschlossen, die bisher zwischen Hagedoorns ursprünglichen Migrationsmethoden (Hagedoorn, 1954) und den auf Grundlage von Schneider (1978) entstandenen analytischen Beschreibungen herrschte.

Weitere Aspekte der (Kirchhoff-)Migration wurden sowohl analytisch als auch geometrisch betrachtet und anhand eines synthetischen Datenbeispiels nummerisch überprüft. Schließlich wurde die Kirchhoff-Migration in einen Arbeitsablauf basierend auf der CRS Stapelung integriert. Diese Kombination aus daten- und modellbasierten Verfahren liefert sehr gute Ergebnisse, wie mit Hilfe eines realen Datenbeispiels aus dem Oberrheingraben gezeigt wurde. Es stellte sich dabei heraus, dass die Verknüpfung von analytischen und geometrischen Betrachtungen äußerst hilfreich ist, um einerseits die Prozesse der seismischen Datenverarbeitung besser zu verstehen und andererseits die erzielten Resultate besser interpretieren und bewerten zu können.

Contents

Chapter 1

Introduction

1.1 Geophysical investigations

The Earth's interior has always been an interesting object for mankind. There are, for example, many natural resources hidden beneath the surface, the Earth's core is the driving force for the magnetic field of our planet, and the activity of the Earth's mantle gives rise to plate tectonics and earthquakes. Unfortunately, the area of interest is not directly accessible for investigations, even if we consider the deepest (and very expensive) boreholes of more than ten kilometers depth—just a tiny scratch compared to the Earth's radius of about 6380 kilometers. Therefore, scientists rely not on direct but on indirect measurements to explore the Earth's interior.

Many geophysical methods and mathematical tools have been introduced and developed in the course of the years to allow the investigation of the Earth's interior. There exist, on the one hand, passive measurements of, e. g., the Earth's gravity or magnetic field, thermal heat flow, or tectonic stress and activity (earthquake seismology). On the other hand, there are active measurements that include the use of an artificial source of energy. Among these are, e. g., geoelectricity, or refraction and reflection seismics. All of these methods are sensitive to different features of the Earth, like mass distribution, elastic properties, thermal and electrical conductivity, just to name a few. With the help of the underlying physical theory and by proper processing and interpretation, the measured data at the surface allow to infer information on the structure and the properties of the Earth's interior. This is the goal of most geophysical activities that are concerned with the solid Earth.

In the following, I will focus on the reflection seismic method. While this method is mainly applied in the oil and gas industry in the search for hydrocarbon reservoirs, the theory and methods presented in this thesis are not restricted to exploration geophysics or even geophysics in general. It is only a question of the length scale of the targets under investigation. In global geophysics, scientists are interested in exploring the entire volume of the Earth, the scale of which is several thousands of kilometers. Exploration geophysicists usually deal with length scales of several tens of kilometers in order to identify targets of a few tens or hundreds of meters within the respective subsurface volume. Environmental or engineering geophysicists conduct surveys with an average length of only a few tens or hundreds of meters and target sizes of about a meter or less. In medicine or materials science (e. g., non-destructive testing applications), the length scales are even smaller, ranging from the centimeter to the submillimeter scale. However, all these areas of operation have something in common: the governing equation for the problem is some form of the wave equation (either acoustic, elastic, or electromagnetic) and the signal wavelengths used in the measurements are small compared to the characteristic length scale of the object under investigation. These aspects allow to apply the theory, methods, and algorithms presented below to a variety of different problems in different working areas (Fink et al., 2002).

1.2 Reflection seismics

The first seismic experiments were performed in the middle of the 19th century. The method proved to be very successful and led to first oilfield discoveries in the 1920's in Oklahoma. At that time, the equipment as well as the processing and interpretation tools were very limited compared to the state-of-the-art that is available to the hydrocarbon industry and academia these days. Although many technical aspects have changed during the years (mainly caused by the rapid development and the increase of computing power), the principle of the reflection seismic method still holds: a seismic source, e. g., a sledgehammer, an explosion, a vibrator truck, or an air gun, emanates elastic or acoustic waves into the Earth. The Earth's response is then recorded by a number of receivers, called geophones on land or hydrophones on sea. Such an experiment is repeated several times for different source locations in order to produce a so-called multicoverage dataset that is the input for seismic data processing.

Unfortunately, the raw multicoverage dataset measured in the field can often hardly be interpreted due to its complexity. There are reflections from discontinuities in material properties, multiple reflections within layers, refracted waves, guided waves, head waves, surface waves which are exponentially attenuated with depth, mode-converted waves, and many others. In addition, amplitude and phase changes occur with respect to the seismic signal that was generated at the source. Figure 1.1 shows an example of some of the complexity of a seismic shot record. Although all waves that have traveled through the Earth contain information about the subsurface, only a small subset is actually being investigated in reflection seismics: primary reflection events, i. e., recorded waves that were reflected only once on their way through the Earth. All other recorded events are treated as noise and geophysicists undertake a lot of efforts to suppress them during the data processing.

All indirect measurements reflect integral properties of the solid Earth. However, it is desirable to derive local properties from these measurements. In order to obtain information about the lithology and local physical properties of the subsurface from the seismic reflection dataset, all propagation effects that influenced the recorded waves on their way through the Earth (Sheriff, 1975) have to be taken into account and should, in principle, be undone. That is the goal of seismic processing and imaging: to transform seismic data measured at the Earth's surface into an image of some property of the Earth—usually wave propagation velocity or acoustic or elastic impedance contrasts—which may then be subject to further analyses.

1.3 Seismic modeling vs. seismic migration/inversion

As was mentioned in the previous section, the goal of seismic migration or inversion is to estimate the subsurface structure and medium parameters from observations at the measurement surface, usually the Earth's surface. Such inversion processes are closely related to forward modeling. Seismic forward modeling is a direct problem: all material properties that define the target of investigation are given.

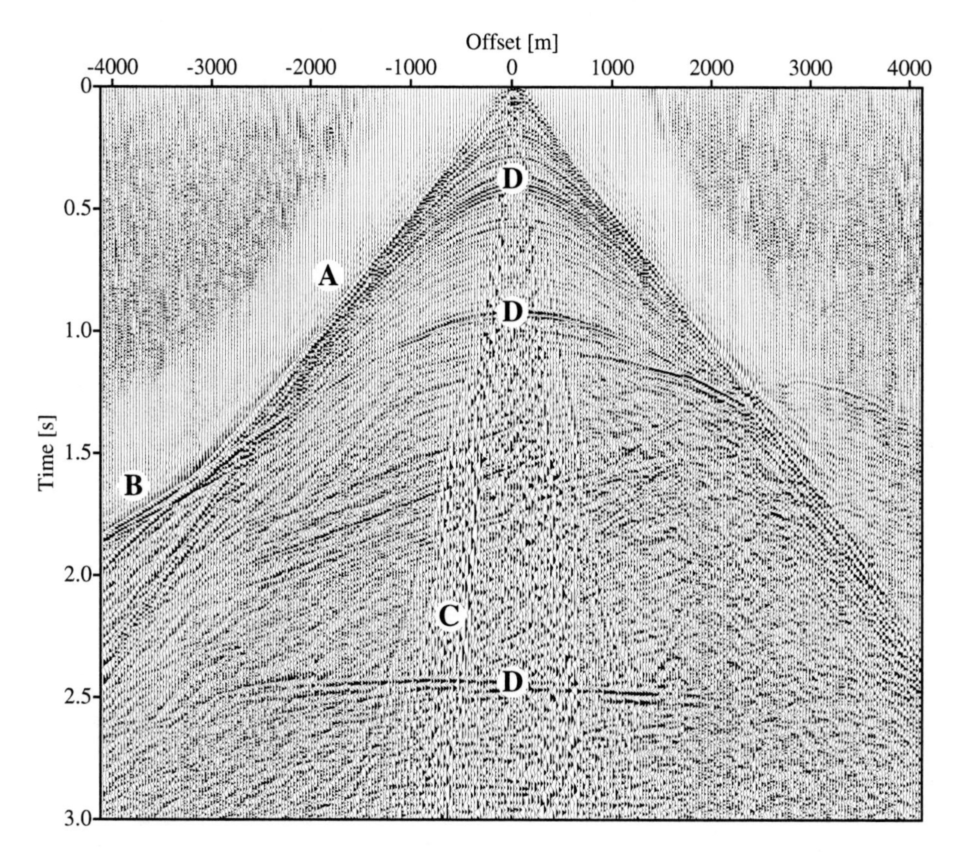

Figure 1.1: A shot record extracted from a multicoverage land seismic dataset. Each trace was recorded by a geophone and is now displayed according to the offset, i. e., the shot-receiver distance. The recorded wavefield is quite complex: Here, the direct wave (A), a head wave (B), ground roll (C), i. e., surface waves traveling at low velocity along the measurement surface, and some primary reflections (D) can be readily observed. In addition, many other types of waves that are difficult to classify were recorded. Moreover, the whole section contains background noise that might obscure actual reflection events. Note that the amplitudes in this section are scaled for display in order to balance the amplitudes of different events.

Furthermore, all initial conditions (i. e., the state of the wavefield in the beginning) as well as all boundary conditions (i. e., the behavior of the wavefield at boundaries) are given and well defined. Then, one can determine the wavefield at any place at any time with the help of the relevant physical equations. On the other hand, an indirect or inverse problem in the context of seismics is more difficult to handle: the wavefield is observed in a finite region at the boundary surface for a certain time window. Then, the aim is to determine all relevant material properties and their discontinuities that define the model. In the field of seismic exploration, such an inverse problem is called migration or imaging. Unfortunately, inverse problems are often ill-posed and suffer from ambiguous solutions—geophysical inverse problems are no exception. Therefore, we have to make some assumptions on the underlying model or we have to simplify the problem in order to obtain a reasonable and unique solution.

There are mainly two types of spatial variations within the Earth. Firstly, there are continuous or so-called smooth changes of material properties. Secondly, there are abrupt changes of parameters, mostly in the vertical direction. While the gradual changes have only a minor effect on traveling waves and cause, e. g., raypaths associated with the waves to be bent, the sharp changes give rise to reflections, refractions, and diffractions that may be recorded at the surface. If the Earth had only gradual changes, we would hardly be able to recognize any reflection events in seismic data acquired along a profile of typical length. This actually means that the reflection seismic method is sensitive to sharp changes in material properties (e. g., wave propagation velocities or densities) of the Earth. Usually, we have some knowledge about the smoothly varying part of the material properties (a so-called macro model) that can be derived from the measured data and we try to find the sharp discontinuities by a suitable imaging procedure called migration.

1.4 Seismic migration

In the early days of seismic exploration, the applied imaging procedures were quite simple. From field records, reflection events at zero-offset, i. e., a coinciding source and receiver location, were extrapolated. The obtained section was then turned upside down, i. e., the time axis pointed downwards. Afterwards, the depth of a reflector directly under the source location was estimated from the zero-offset reflection events by appropriately scaling the time axis. However, it was soon discovered that this technique leads to inaccurate or poor images if reflectors in the subsurface are dipping—even though the velocity might have been correctly determined. This is due to the fact that the directions from which waves arrived at a receiver were not taken into account.

It was Hagedoorn (1954) who developed a technique to overcome the above-mentioned problems and to move reflection events to their correct spatial location, given a correct velocity model. If a circle is drawn around the origin of each trace within the approximated zero-offset section with a radius that is simply determined by half the arrival time of a reflection event times the estimated propagation velocity, one observes that all these circles have a common envelope. This envelope defines the correct spatial location of the reflector. This approach does not only scale the arrival time of reflection events but laterally moves the events—the concept of migration was born. As there was no additional expenditure required and the technique could be readily applied even in the field, it has meanwhile also become known as *shoestring migration* (Bleistein, 1999), see also Figure 1.2. Admittedly, the concept by Hagedoorn contains many (over)simplifications, but it represents the fundamental idea of migration. Each of the circles corresponds to an isochron, i. e., the line of equal reflection time. Of course, the isochrons are only half-circles (2D) or half-spheres (3D) in the case of a constant wave

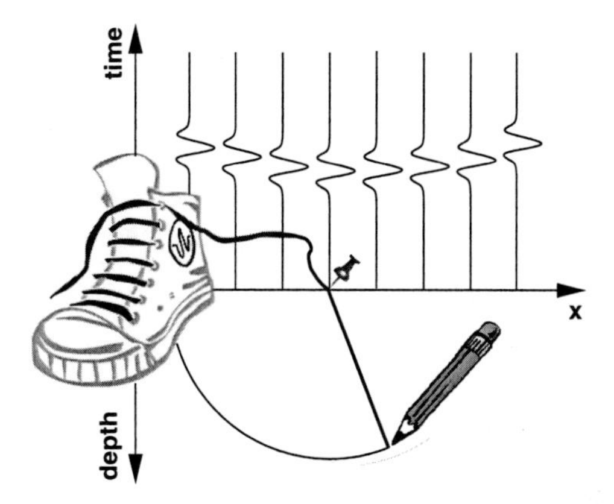

Figure 1.2: The principle of migration developed by Hagedoorn (1954), also known as migration on a shoestring. For a reflection event in the time domain, a correctly scaled half-circle is drawn around the origin of each trace. A reflector is defined by the envelope of all these half-circles.

propagation velocity in the subsurface. For inhomogeneous media, the shoestring can no longer be used and has to be replaced by more sophisticated arcs.

The same results as in Hagedoorn's migration can be obtained if diffraction traveltime curves or surfaces are used instead of isochrons. The diffraction traveltime is obtained if a single point diffractor is illuminated by the seismic experiment under consideration. Then, a reflection event in the time domain is not spread out along an isochron, but amplitudes of reflection events are summed up along the diffraction traveltime curve/surface and assigned to the corresponding depth point. The close relationship between the isochron and the diffraction traveltime curve/surface on the one hand, and the reflector in depth and the reflection event in the time domain on the other hand, has become known as the concept of duality (Tygel et al., 1995). In seismic exploration, the above-mentioned migration method of using diffraction traveltimes is called *diffraction-stack migration*. It was later related to the wave equation and became familiar as *Kirchhoff migration* (Schneider, 1978). The name was chosen with regard to the Kirchhoff integral which is used to describe the (forward) propagation of seismic waves within a given depth model. As the Kirchhoff integral itself cannot be used to solve the inverse problem, i. e., to describe backward propagation, Kirchhoff migration was introduced as its adjoint operation that is based on the forward propagation of the recorded wavefield in the reverse (time) direction.

The actual objective of seismic migration is quite simple. The migration process should move reflection events to their proper spatial location, or in other words, "migration is the art of reversing wave propagation effects in order to obtain clear images of the subsurface" (M. Landrø). This implies that

- the position and the dip of reflection events are corrected in order to coincide with the real reflector positions (Figure 1.3a)),
- diffractions are collapsed into the corresponding diffraction point (Figure 1.3b)),
- triplications are resolved (Figures 1.3c) and d)),
- the influence of the topography of the measurement surface is removed, and

- geometrical spreading effects are removed if a true-amplitude migration is applied. The term "true-amplitude" is described in detail in the next section.

In addition, the spatial resolution is (in many cases) improved as energy gets focused in the migration process, i e., the reflector Fresnel zone in the depth domain is usually smaller than the projected Fresnel zone at the measurement surface. Additional comments on the Fresnel zone are given in Chapter 6.

1.5 True-amplitude migration

Traditional Kirchhoff migration considers traveltime to be the only important parameter for imaging. The reflector is imaged in the sense that its position and shape are correctly represented. However, there is no attempt to recover information about the material parameters of the subsurface. That is why one had to distinguish between migration and inversion in the past. In the mind of early geophysicists, the output of a migration procedure was just a processed section and a structural image of the subsurface, as opposed to the inversion result that should provide a subsurface parameter image. In recent years, the distinction between migration and inversion has blurred as modern migration techniques do attempt to handle not only kinematic (i. e., traveltime-related) but also dynamic (i. e., amplitude-related) information. This is a consequence of advances made in the 1970's, when the technique of identifying gas-bearing strata by apparent high-amplitude bright spots on seismic sections was established. Today, the correct handling of amplitudes has become an ever increasing subject in seismic imaging, inversion, and reservoir characterization.

If one talks about physically well-defined amplitudes in migrated seismic sections, one immediately arrives at the term *true-amplitude*. A very clear definition of a "true-amplitude reflection" was given by Newman (1973). The definition could not be simpler: the true-amplitude reflection is the reflected seismic signal, from which the spherical divergence effect (also called geometrical spreading) is removed. The energy in body waves spreads out as the spherical wavefront moves away from the source and thereby usually expands, causing the energy density to vary, which is above-mentioned effect. According to Newman the removal of geometrical spreading effects is necessary, or in his words a "compensation for these effects is mandatory, if reflection amplitudes are to be of diagnostic value". It should, however, be indicated that the term "true-amplitude" was not used in his paper and the origin of the word remains unknown. Newman's contribution to true amplitudes is much more profound than what is indicated here. He can also be considered as the inventor of true-amplitude migration, see Newman (1975). In a true-amplitude migration process, the geometrical spreading effect is removed from the input seismograms by suitable weight functions and the resulting amplitudes in the migrated section become a measure of the angle-dependent reflection coefficient. This is the basis for studies of the amplitude variation with offset (AVO) or angle (AVA) in order to determine specific reservoir characteristics. Migration is now routinely applied and has become a central part in seismic imaging (Figure 1.4) while it was previously often only an optional or the final step in seismic data processing.

1.6 High-frequency asymptotics

The idea of the solution of the inverse problem that was outlined above is based on a geometrical optics approach, usually called seismic ray theory in this context. In fact, it is difficult to comprehend

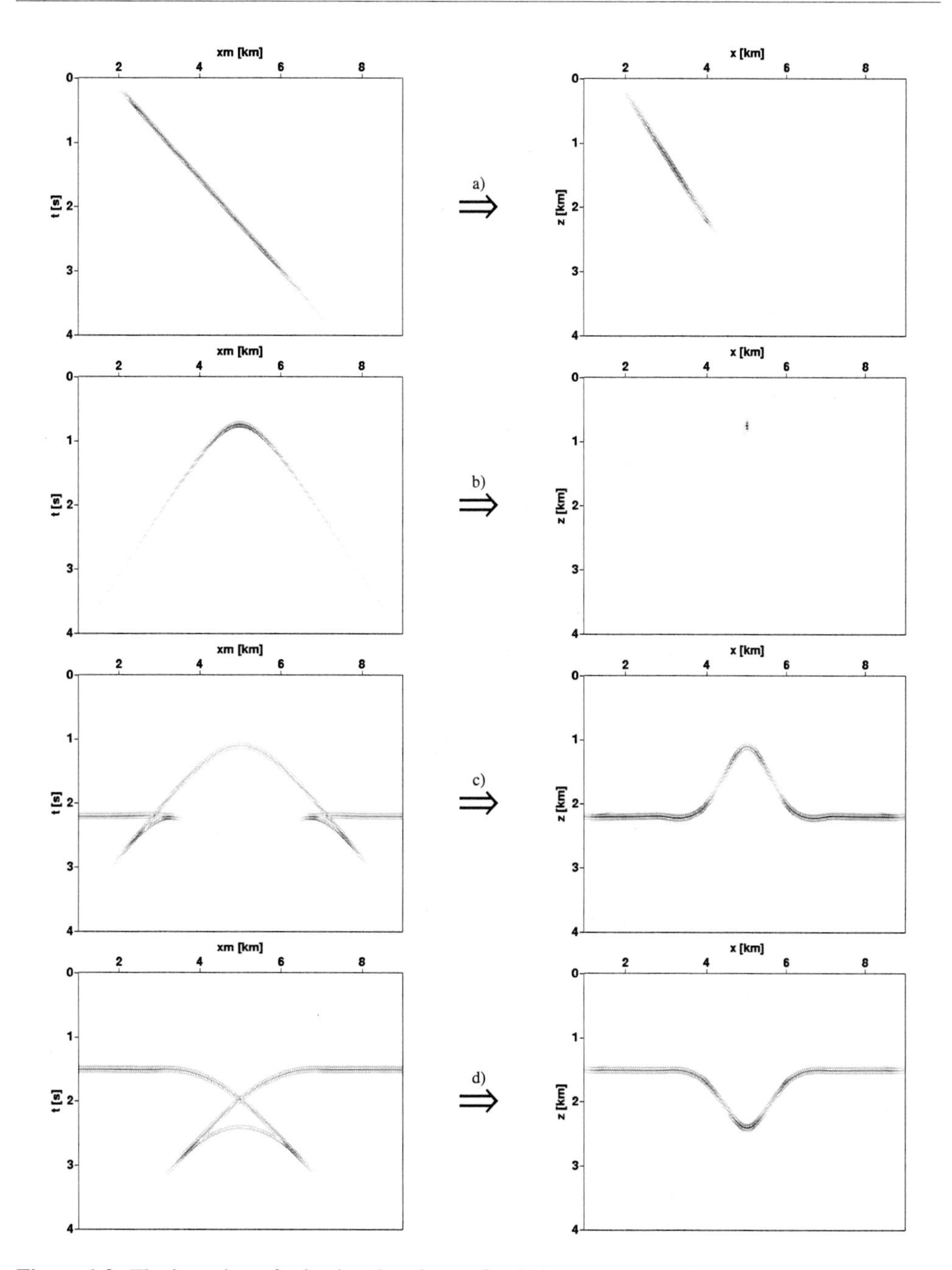

Figure 1.3: The intention of seismic migration: reflection events in the time domain are transformed into images of the corresponding structures in the subsurface (left column: zero-offset sections; x_m is the midpoint location which in the case of zero offset is identical to the source position; right column: depth-migrated sections). It can easily be observed that migration is necessary to obtain a clear and correct image of the subsurface.

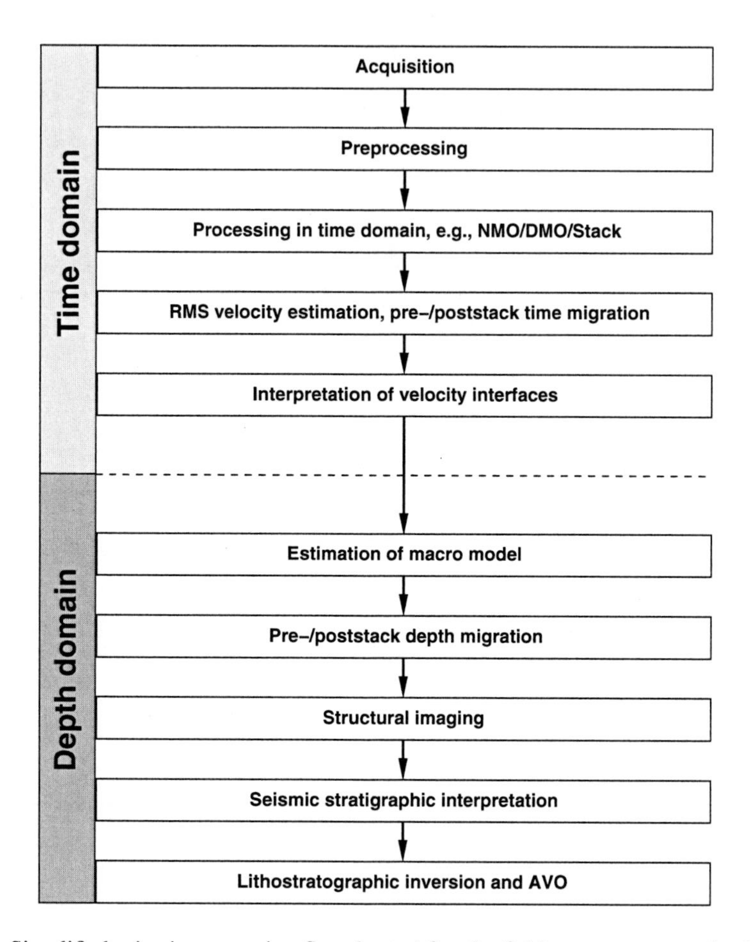

Figure 1.4: Simplified seismic processing flowchart. After the field measurements, the data are preprocessed and processed in the time domain, e. g., by the classical normal moveout (NMO)/dip moveout (DMO)/stack method. A root-mean-square (RMS) velocity model is necessary in order to perform a time migration. For a depth migration, a macro Earth model (interval velocities) has to be estimated. The depth-migrated images serve as input for structural as well as stratigraphic interpretation and lithostratigraphic and amplitude-variation-with-offset (AVO) analyses. Thus, the migration process has become a central part in seismic imaging. [based on Farmer et al. (1993)]

the concept of true amplitudes without some understanding of ray theory. According to this theory one can decompose the wavefield recorded in a seismic trace into a number of individual events and assign to each one a reflection time and a signal or pulse. Its amplitude depends, in principle, on a source and receiver factor, a transmission and attenuation factor, a reflection factor, and a geometrical spreading factor. Without this ray theoretical description of a seismic reflection one cannot understand the indicated factors nor can a true-amplitude reflection be defined.

Much has been accomplished by assuming that the rules for the reflection of waves at reflectors in the real Earth can be locally approximated by the reflection of plane waves from plane reflectors between constant-velocity acoustic media. These simple ideas may be applied to heterogeneous media with curved reflectors only when the waves are of sufficiently high frequency. As is outlined in Chapter 3, the term "high frequency" does not refer to absolute values of the frequency content of waves. What must be considered is the relationship between the wavelengths associated with the frequencies in the data and a characteristic length scale of the medium. Under the high-frequency assumption, the propagation of waves can be approximated by the propagation of wave packets along certain paths called rays. The reflection and transmission mechanisms are simply governed by Snell's law. One has to keep in mind that these principles might not only be considered in forward modeling, but also in inversion. A theoretical description of the inversion process involves the Green's function of the medium under consideration. Physically, the Green's function describes the (non-trivial) response of a general linear system to a unit force represented by a Dirac delta function[1]. This Green's function is, in general, unknown for realistic Earth models and can only be calculated exactly for simple types of media. Therefore, we need an approximation of the Green's function, and it has turned out that in many cases ray theory is able to provide such an approximation. We have also seen in earlier sections that Kirchhoff migration is based on the ideas of wavefronts, reflectors, and raypaths—all of them concepts of high-frequency asymptotic wave theory. These principles are used throughout this thesis.

In mathematical terms, we can describe one reflection event within a seismic trace by

$$U(t) = R_c \, \frac{\mathcal{A}}{\mathcal{L}} \, F(t - \tau_R) \,, \tag{1.1}$$

where R_c denotes the angle-dependent (plane-wave) reflection coefficient, $\mathcal{L}$ symbolizes the point-source geometrical spreading factor, and $\mathcal{A}$ represents all other effects on the amplitude, such as source strength, source and receiver coupling, transmission loss, and attenuation in the reflector's overburden, to name but a few. Moreover, $F(t)$ is the analytic (i. e., complex) source wavelet which is shifted to the arrival time τ_R (reflection traveltime). A seismic trace with several (primary) events may be described by a superposition of individual seismic events of the type of equation (1.1). A complex description is used in order to handle phase shifts along the raypaths.

1.7 Source-receiver configurations

The simple migration scheme that was outlined above is only applicable for data where sources and receivers are located at the same point on the Earth's surface.[2] However, this is a very special situation and we need to have a look at other frequently used source-receiver configurations in seismic data processing. Among these are

[1]The Dirac delta function is, in fact, a distribution and belongs to the class of generalized functions. Nevertheless, it is rather called delta function than delta distribution in the literature.

[2]Note that Hagedoorn (1954) did not restrict his considerations to this simple case.

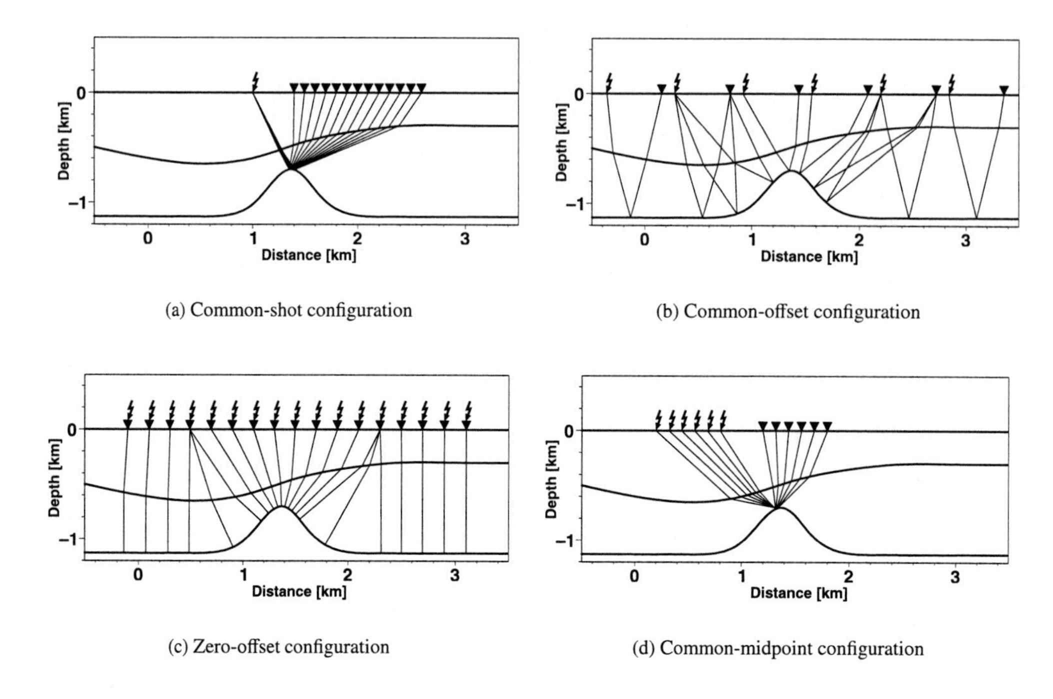

(a) Common-shot configuration

(b) Common-offset configuration

(c) Zero-offset configuration

(d) Common-midpoint configuration

Figure 1.5: Shot-receiver configurations. Shots are marked by lightnings while receivers are marked by triangles. In addition, associated raypaths are plotted. (a) Common-shot (CS) configuration. (b) Common-offset (CO) configuration. (c) Zero-offset (ZO) configuration. (d) Common-midpoint (CMP) configuration. [based on Höcht (2002)]

- the common-shot (CS) configuration. This is the usual field measurement configuration. A CS gather consists of traces recorded at many receiver positions with increasing distance (offset) from a single source.
- the common-offset (CO) configuration. A CO gather consists of a collection of traces whose source-receiver distance (offset) is constant.
- the zero-offset (ZO) configuration. A ZO gather is a special CO gather where the offset equals zero, i. e., source and receiver locations coincide for each trace.
- the common-midpoint (CMP) configuration. A CMP gather refers to a collection of traces recorded with different offsets but with the same position on the measurement surface being the midpoint of the source-receiver pairs. The CMP geometry is important as it provides a well-established collection of data that can be used to construct an approximate zero-offset profile. In the literature, the CMP gather is also often called common-depth-point (CDP) gather. However, one should note that CMP and CDP gathers are, in general, only identical if reflectors in the subsurface are horizontal and there are no lateral variations of material parameters.

Note that only a CS gather consists of traces that can be recorded by a single seismic experiment. All other gathers contain traces combined from different seismic experiments. Usually, CO and CMP gathers are extracted from the multicoverage dataset that was recorded in the field by resorting the seismic traces. A strict ZO gather cannot be extracted from a multicoverage dataset as the associated acquisition configuration cannot be utilized in the field because of high amplitude reverberations associated with typical seismic sources such as explosions, vibrators, or air guns. A zero-offset gather has to be simulated by suitable imaging methods.

Throughout this thesis, I assume that the measurement surface is densely covered with source-receiver pairs (S,R). In order to describe the location of a seismic trace in dependence on a shot and receiver position as well as the recording geometry, equation (1.1) has to be modified. U becomes a function of S and R, i. e.,

$$U = U(S,R,t) = U(S(\vec{\xi}),R(\vec{\xi}),t) = U(\vec{\xi},t)\,, \tag{1.2}$$

where $\vec{\xi}$ is the configuration vector that varies in the plane $z = 0$. The 2D position vectors in the plane $z = 0$ of sources and receivers can be expressed as

$$\vec{r}_S(\vec{\xi}) = \vec{r}_{S0} + \mathbf{\Gamma}_S\,\vec{\xi} \quad \text{and} \quad \vec{r}_R(\vec{\xi}) = \vec{r}_{R0} + \mathbf{\Gamma}_R\,\vec{\xi}\,, \tag{1.3}$$

where $\mathbf{\Gamma}_S$ and $\mathbf{\Gamma}_R$ are $2{\times}2$ matrices which account for the different source-receiver configurations. The vectors $\vec{r}_{S0}$ and $\vec{r}_{R0}$ depend on the choice of the origin of $\vec{\xi}$. In other words, the vector $\vec{\xi}$ denotes a trace position whereas the matrices $\mathbf{\Gamma}$ describe the measurement configuration. For a regular common-shot configuration, the configuration matrices are given by $\mathbf{\Gamma}_S = \mathbf{0}$ and $\mathbf{\Gamma}_R = \mathbf{I}$, where $\mathbf{0}$ and $\mathbf{I}$ are the 2×2 zero and unit matrices, respectively. For a regular common-offset configuration, $\mathbf{\Gamma}_S = \mathbf{I}$ and $\mathbf{\Gamma}_R = \mathbf{I}$. Details about this parameterization can be found in Schleicher (1993). The region A in the plane $z = 0$ over which the vector $\vec{\xi}$ varies to cover all source-receiver pairs is denoted the aperture of the seismic experiment, see Figure 1.6.

If data are acquired along a surface with topographic variations, the description of actual source and receiver positions becomes more complicated. However, to describe the lateral positions it is still sufficient to use 2D position vectors in the plane $z = 0$. In addition, we need to know the surface elevation at both the source and the receiver location. In principle, one can choose the sources and receivers on the measurement surface such that their projections are equally spaced in the plane $z = 0$. Then, the configuration matrices remain unchanged, at least as long as only the recording geometry is considered. The projection itself is a one-to-one relation for all practical cases. However, in reality the regular geometry is often chosen to be along the curved measurement surface and, thus, we have an irregular spacing of the source and receiver projections in the plane $z = 0$. As a consequence, we need some kind of binning for all practical cases and the configuration matrices are no longer constant but dependent on the configuration vector $\vec{\xi}$. For the actual determination of the true-amplitude migration weight function, the configuration matrices need to be further extended to account for the local dip of the topography (Schleicher et al., 1993; Spinner, 2003).

1.8 Migration methods

There exist numerous migration methods which can roughly be divided into methods based on

a) an integral solution (Kirchhoff migration),

b) a derivative solution (finite-difference wave equation migration), or

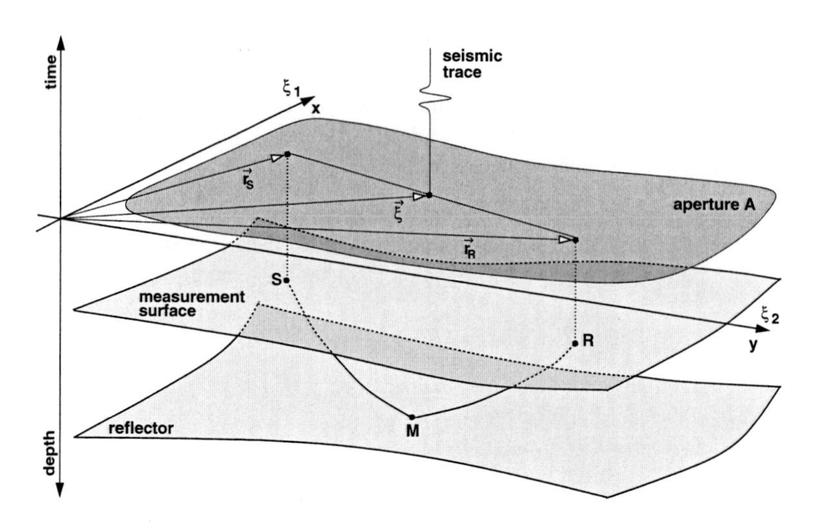

Figure 1.6: Seismic recording geometry. An event in the seismic trace recorded at R is caused by a wave that emanates from the source S and reflects at the depth (reflector) point M. The actual positions of the source and receiver on the measurement surface are vertically projected into the plane $z = 0$ to obtain the 2D position vectors $\vec{r}_S$ and $\vec{r}_R$, respectively. The location of a seismic trace can then be described by the configuration vector $\vec{\xi}$ in combination with the configuration matrices $\mathbf{\Gamma}$. The aperture A in the plane $z = 0$ of the seismic experiment is the area over which $\vec{\xi}$ varies to cover all source-receiver pairs.

c) a Fourier domain solution (frequency-wavenumber migration)

of the wave equation. In addition, we have to distinguish between 2D and 3D, time and depth, and prestack and poststack migration. A general overview of migration methods is given in Yilmaz (2001), advantages and disadvantages of each method are discussed in, e. g., Gray et al. (2001). A suitable macro Earth model is always needed in order to migrate the data. I assume that such a model (or, at least, an initial macro model) is available, see also Chapter 7 for details on estimating a velocity model. In this thesis, I restrict my considerations to Kirchhoff migration. It has an intuitive geometrical interpretation and the derivation of true-amplitude weights is—as opposed to other methods—straightforward for most shot-receiver configurations.

While poststack migration takes a simulated zero-offset section (stacked section) as input, prestack migration handles the complete preprocessed multicoverage dataset. Thus, poststack migration is much more efficient than prestack migration but fails if the subsurface structure is complex. This is not an inherent problem of poststack migration but due to the underlying assumptions that are commonly made to simulate the stacked time section. In addition, we must remember that ZO rays might illuminate less regions in the subsurface than the rays associated with the multicoverage dataset. Prestack migration does not suffer from such problems, it is more time consuming, though. Here, each common-offset panel is migrated separately—this allows to analyze common-image gathers (CIG) to control and update the velocity model. A CIG is formed by collecting all those traces after migration that represent the same lateral position but belong to different offset panels. As a consequence, prestack migration is often applied in an iterative way in order to find the best velocity model that is consistent with the data. Figure 1.7 shows the different approaches for pre- and poststack migration.

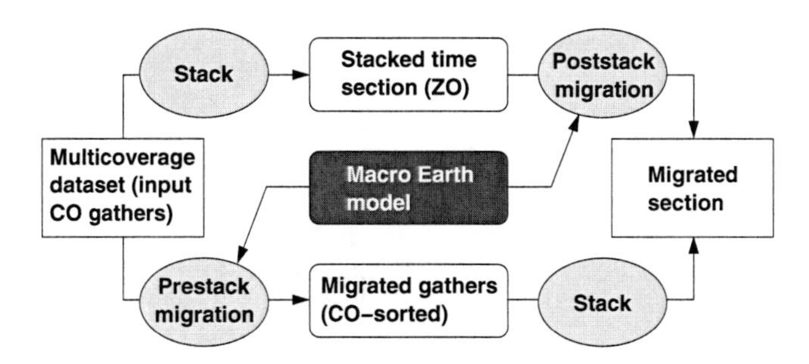

Figure 1.7: Prestack versus poststack migration. While poststack migration takes a simulated ZO section as input, prestack migration deals with the complete multicoverage dataset. In other words, in prestack migration each CO panel is migrated separately. Note that stacking in CIGs after prestack migration averages the amplitudes. As a consequence, the final prestack-migrated section is no longer true-amplitude, even if all individual offset gathers were migrated in such a way. [based on Fagin (1999)]

Both, the prestack and poststack Kirchhoff migration operations can be performed as time or depth migration, respectively. While the output of a time migration is an image in time, depth migration produces a real subsurface image in the depth domain which seems at first to be more favorable. However, Kirchhoff time migration has some advantages: it is a very fast process based on straight rays and we need only an integral quantity of the subsurface to perform the imaging step, namely a root-mean-square (RMS) velocity model. For this reason, time migration is less sensitive to errors in the velocity model compared to depth migration where interval velocity models and, therewith, local quantities are necessary. However, time migrations fail as soon as the subsurface has (strong) lateral inhomogeneities because the assumptions of straight rays and an RMS velocity field are no longer valid. Figure 1.8 shows the ranges of applications for pre- and poststack time and depth migration in dependence on the structural complexity and lateral inhomogeneity of the subsurface.

The ultimate but most challenging task is 3D prestack depth migration. Such processes are increasingly applied in the oil and gas industry today. Hence, it is not surprising that one will find the most powerful supercomputers, e. g., clusters with thousands of CPUs, in associated companies. These companies undertake great efforts to develop imaging algorithms that do not only run faster compared to existing standards but also produce images of higher quality. In addition, they develop many tools to extract as much detailed information as possible from the data. Here is an example to demonstrate the huge amount of operations necessary in processing prestack data. Assume there is an aperture of about 2000 m on either side of a central survey line in a seismic reflection experiment. With a typical bin width of 12.5 m, this means 320 bin-lines contribute to the Kirchhoff summation process. In total, and for a single offset, a single subsurface point will be imaged by a weighted sum of over 100,000 different samples. Supposing a nominal fold[3] coverage of 50, the total number of samples contributing to the subsurface image of this sole point is more than five million. To find each of these samples, we have to calculate the traveltimes and weights beforehand, i. e., the Green's functions need to be estimated. Then a weighted sum is performed to obtain the migration output value. And yet, only a few tens or hundreds of samples, those corresponding to an aperture equal to the projected Fresnel zone around the specular ray, contribute to the summation and give the amplitude of an image point. While for a 2.5D Kirchhoff migration problem with a seismic line of about 10 km length there are

[3] A fold is the multiplicity of common-midpoint data.

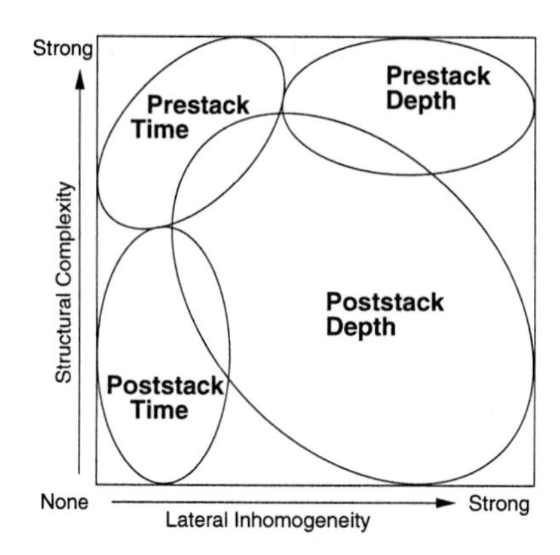

Figure 1.8: The range of application for pre- and poststack time and depth migration. If there are lateral inhomogeneities in the subsurface, depth migration has to be performed in order to obtain a correct Earth image. Prestack migration is necessary if the subsurface shows structural complexity. [based on Liner (1999)]

around a million of these points which must be considered, in 3D this means in order to be able to image a 500 km^3 subsurface volume, we must image about 4.8 billion points (Robein, 2003). This gives an idea of the huge quantity of calculations required in prestack depth migration and why this technique has remained reserved to solving specific problems. This will, however, become less the case as computing power continues to increase.

1.9 Outline of this thesis

In this chapter, I gave a brief introduction to reflection seismics and seismic reflection imaging. Chapter 2 is devoted to the wave equation that is the governing equation of forward and inverse problems in seismics. Because of this fundamental importance, I present the ideas leading to the elastic and acoustic wave equation, respectively. In Chapter 3 a high-frequency asymptotic method based on the wave equation is reviewed. I derive the basic equations of this so-called ray method, the eikonal and transport equation, and show how to obtain their solutions. The general forward and inverse problems are dealt with in Chapter 4. Starting from the wave equation, the Kirchhoff integral (that is frequently used in forward modeling) and the Porter-Bojarski integral (that is frequently used in inversion) are derived and the mathematics underlying the process of migration are presented. In Chapter 5 a geometrically more appealing way to describe the migration process is shown along with suitable weighting functions that cause the migration result to be true-amplitude. Based on the described geometrical concepts, migration artifacts are explained in Chapter 6, both mathematically and geometrically. For the first time, a geometrical interpretation of the stationary-phase approximation of the migration integral in relation to the diffraction and reflection traveltime surfaces in the time domain is presented. In Chapter 7 I address further aspects of (Kirchhoff) migration, mainly related to practical problems and their solutions, e. g., aliasing and anti-aliasing filters. A synthetic data example is introduced in Chapter 8. By means of a complex model, the aspects of true-amplitude Kirchhoff migration that are explained in previous chapters are demonstrated and examined both qualitatively and quantitatively. Finally, I present the migration process in the context of a data-driven imaging workflow in Chapter 9.

The potential of my approach is demonstrated on a real data example. The conclusions are given in Chapter 10. Appendix A deals with the 1D Method of Stationary Phase, and the projected Fresnel zone is explained in Appendix B.

Chapter 2

The wave equation

The governing equation for seismic reflection modeling as well as migration/inversion is some form of the wave equation. Because of their fundamental importance, I give a brief overview of the ideas leading to the elastic and acoustic wave equations, respectively. Detailed and comprehensible explanations can be found in, e. g., Aki and Richards (1980), Smith (1993), Lay and Wallace (1995), Scales (1997), or Červený (2001). See also references therein.

2.1 The elastodynamic wave equation for an anisotropic inhomogeneous unbounded medium

We start with a look at a subvolume within a solid body. Assuming that this small particle has a volume V and a boundary surface S, we can write Newton's second law of mechanics in the form

$$\iiint\limits_V \rho \frac{\partial^2 \vec{u}}{\partial t^2}\, dV = \iiint\limits_V \vec{f}\, dV + \oint\!\!\!\oint\limits_S \vec{T}(\vec{n})\, dS \;, \tag{2.1}$$

where $\vec{u}(\vec{x}, t)$ is the displacement vector, t is time, and $\rho(\vec{x})$ is the density of the solid body. The density of the external body forces is described by $\vec{f}$ and the traction acting on the particle volume V through the boundary surface S is given by $\vec{T}(\vec{n})$. The vector $\vec{n}$ is a unit vector normal to the boundary surface and points outwards. By converting the surface integral into a volume integral by means of Gauss' theorem, we can rewrite equation (2.1) in the differential form

$$\rho \frac{\partial^2 u_j}{\partial t^2} = f_j + \frac{\partial \tau_{ij}}{\partial x_i} \;, \qquad i, j = 1, 2, 3, \tag{2.2}$$

where x_i are the Cartesian components of the position vector $\vec{x}$, and summation over repeated indices is implied. The tensor $\underline{\tau}$ is called (Cauchy) stress tensor; it is a symmetric tensor of second order, i. e., $\tau_{ij} = \tau_{ji}$, describing the stress condition at any point $\vec{x}$ in the medium. The diagonal terms of the stress tensor, τ_{ii}, are called normal stresses while the off-diagonal terms are called shear stresses. The set of three equations (2.2) is called *equation of motion* for a continuum. These equations relate the density-weighted accelerations to body forces and stress gradients in the medium and form, thus, the most fundamental equations underlying the theory of seismology as they relate forces in the medium to measurable displacements.

Under the assumptions of small deformations and an anisotropic perfectly linear-elastic solid, each component of the stress tensor, τ_{ij}, represents a linear combination of the components of the strain tensor, e_{kl}, i. e.,

$$\tau_{ij} = c_{ijkl}\, e_{kl} \,. \tag{2.3}$$

The parameter $\underline{\underline{c}}$ is the stiffness tensor, sometimes also called tensor of elastic parameters or simply elastic tensor. The constants of proportionality, c_{ijkl}, in the stress-strain relationship are known as elastic moduli and define the material properties of the medium. The strain tensor $\underline{e}$ is assembled by spatial gradients of the displacement field, i. e., the strain at a point describes the deformation in the vicinity of that point. Equation (2.3) is a generalized version of Hooke's law that assumes a generally anisotropic linear-elastic medium. This is the only constitutive relation between stress and strain we will consider here. To describe nonlinear-elastic solids, one would have to add higher order terms to the right-hand side of equation (2.3), i. e., $\tau_{ij} = c_{ijkl}\, e_{kl} + O\left(\underline{e}^2\right)$. Due to the symmetries in the stress and strain tensors and also from thermodynamic considerations, the fourth order tensor $\underline{\underline{c}}$ of elastic moduli is restricted to 21 independent components in the most general anisotropic adiabatic case. The presence of material symmetries will further reduce the number of independent moduli.

For small strains and small displacement derivatives, the nine elements of the strain tensor are given by

$$e_{kl} = \frac{1}{2}\left(\frac{\partial u_k}{\partial x_l} + \frac{\partial u_l}{\partial x_k}\right) . \tag{2.4}$$

Note that the strain components depend linearly on derivatives of the displacement components; higher orders terms may be neglected due to the above-mentioned assumptions. By inserting the relationship (2.4) into (2.3), we obtain for the spatial derivatives of the stress tensor components

$$\frac{\partial \tau_{ij}}{\partial x_i} = \frac{1}{2}\frac{\partial}{\partial x_i}\left(c_{ijkl}\frac{\partial u_k}{\partial x_l} + c_{ijkl}\frac{\partial u_l}{\partial x_k}\right) , \tag{2.5}$$

which can be simplified to read

$$\frac{\partial \tau_{ij}}{\partial x_i} = \frac{\partial}{\partial x_i}\left(c_{ijkl}\frac{\partial u_k}{\partial x_l}\right) \tag{2.6}$$

due to the symmetry property $c_{ijkl} = c_{ijlk}$ of the stiffness tensor. Remember that summation over repeated indices is implied. Substituting this result in equation (2.2) leads to

$$\rho\frac{\partial^2 u_j}{\partial t^2} - \frac{\partial}{\partial x_i}\left(c_{ijkl}\frac{\partial u_k}{\partial x_l}\right) = f_j \,. \tag{2.7}$$

This equation is called the *elastodynamic wave equation for an inhomogeneous anisotropic linear-elastic medium.* It is a hyperbolic system of three partial differential equations of second order and cannot be solved analytically in this general form.

2.2 The elastodynamic wave equation for an isotropic inhomogeneous unbounded medium

Fortunately, the elastic properties for many materials in the Earth are independent of direction or orientation. In this isotropic case, the components of the stiffness tensor, c_{ijkl}, can be expressed in terms of two independent elastic moduli, namely

$$c_{ijkl} = \lambda\, \delta_{ij}\delta_{kl} + \mu\left(\delta_{ik}\delta_{jl} + \delta_{il}\delta_{jk}\right) , \tag{2.8}$$

where δ_{ij} is the Kronecker symbol, i. e.,

$$\delta_{ij} = 1 \text{ for } i = j\,, \qquad \delta_{ij} = 0 \text{ for } i \neq j\,. \tag{2.9}$$

The quantities λ and μ are called Lamé parameters. The significance of the shear modulus, or rigidity, μ, is readily apparent as a measure of resistance to shear stress. With the help of equation (2.8), we can simplify equation (2.7) and write it in vector notation as

$$\rho\frac{\partial^2\vec{u}}{\partial t^2} - (\lambda+\mu)\vec{\nabla}(\vec{\nabla}\cdot\vec{u}) - \mu\Delta\vec{u} - \vec{\nabla}\lambda(\vec{\nabla}\cdot\vec{u}) - \vec{\nabla}\mu\times(\vec{\nabla}\times\vec{u}) - 2(\vec{\nabla}\mu\cdot\vec{\nabla})\vec{u} = \vec{f}\,. \tag{2.10}$$

This equation is now called the *elastodynamic wave equation for an inhomogeneous isotropic linear-elastic medium.* The symbol $\vec{\nabla}$ denotes the Nabla operator, i. e., $\vec{\nabla}\Phi$ is the gradient of a scalar function, $\vec{\nabla}\cdot\vec{\Psi}$ is the divergence of a vector function, and $\vec{\nabla}\times\vec{\Psi}$ is the curl of a vector function, respectively. The symbol $\Delta = \vec{\nabla}\cdot\vec{\nabla}$ denotes the Laplace operator.

2.3 The elastodynamic wave equation for an isotropic homogeneous unbounded medium

If we consider a homogeneous medium, then $\vec{\nabla}\mu = \vec{\nabla}\lambda = \vec{0}$ and equation (2.10) reduces to

$$\rho\frac{\partial^2\vec{u}}{\partial t^2} - (\lambda+\mu)\vec{\nabla}(\vec{\nabla}\cdot\vec{u}) - \mu\Delta\vec{u} = \vec{f}\,. \tag{2.11}$$

This equation is called the *elastodynamic wave equation for a homogeneous isotropic linear-elastic medium.*

2.4 P- and S-waves

As we have seen, the wave equation describes the spatial and temporal development of the displacement field; it admits solutions in the form of traveling waves. Plane waves are regarded as the simplest solution, so we study

$$\vec{u}(\vec{x},t) = \vec{A}\exp\left(i\left[\vec{k}\cdot\vec{x} - \omega t\right]\right) \tag{2.12}$$

in the following. The amplitude $\vec{A}$ is assumed to be constant. The wavenumber vector $\vec{k}$ tells us the direction of propagation; its components are given by k_1, k_2, and k_3, and the absolute value of $\vec{k}$ is the wavenumber $k = |\vec{k}|$. If we neglect body forces and insert the plane wave ansatz into equation (2.11), we will after some manipulations end up with a system of equations in the form

$$\mathbf{N}\vec{A} = \vec{0}\,. \tag{2.13}$$

The matrix $\mathbf{N}$ is given by

$$\mathbf{N} = \begin{bmatrix} (\lambda+\mu)k_1k_1 + \Lambda & (\lambda+\mu)k_1k_2 & (\lambda+\mu)k_1k_3 \\ (\lambda+\mu)k_2k_1 & (\lambda+\mu)k_2k_2 + \Lambda & (\lambda+\mu)k_2k_3 \\ (\lambda+\mu)k_3k_1 & (\lambda+\mu)k_3k_2 & (\lambda+\mu)k_3k_3 + \Lambda \end{bmatrix}, \tag{2.14}$$

where

$$\Lambda = \mu k^2 - \rho\omega^2 \,. \tag{2.15}$$

It is well known that a linear system of homogeneous equations such as (2.13) has a nonzero solution if and only if det(**N**) = 0. This characteristic equation is given by

$$\det(\mathbf{N}) = \Lambda^2 \left(\Lambda + (\lambda + \mu)k^2\right) = 0 \,. \tag{2.16}$$

Thus, there are two possibilities to solve equation (2.16):

1. The solution where

$$\Lambda + (\lambda + \mu)k^2 = (\lambda + 2\mu)k^2 - \rho\omega^2 = 0 \tag{2.17}$$

leads us to so-called *compressional waves* or *P-waves*. The wavenumber is given by

$$k^2 = \frac{\rho\omega^2}{\lambda + 2\mu} \,, \tag{2.18}$$

and due to $k = \omega/v$ we conclude that this wave is associated with the phase velocity

$$v_P = \sqrt{\frac{\lambda + 2\mu}{\rho}} \,. \tag{2.19}$$

The polarization of the wave is given by the eigenvector $\vec{A}_P$ corresponding to the system of equations (2.13) when k^2 satisfies equation (2.18). If we construct this eigenvector, we observe that $\vec{A}_P$ and $\vec{k}$ are collinear, i. e., the polarization of the compressional wave coincides with its direction of propagation. Therefore, such type of wave motion is also called longitudinal.

2. There is an alternative type of elastic wave motion which arises if

$$\Lambda^2 = \left(\mu k^2 - \rho\omega^2\right)^2 = 0 \,. \tag{2.20}$$

Here, the wavenumber is given by

$$k^2 = \frac{\rho\omega^2}{\mu} \,, \tag{2.21}$$

and, once again, we conclude that the corresponding so-called *shear wave* or *S-wave* is associated with the phase velocity

$$v_S = \sqrt{\frac{\mu}{\rho}} < v_P \,. \tag{2.22}$$

There are two mutually orthogonal eigenvectors $\vec{A}_S$ corresponding to the system of equations (2.13) when k^2 satisfies equation (2.21) due to the matrix **N** being symmetric. Both eigenvectors are orthogonal to the direction of propagation, $\vec{k}$, thus such type of wave motion is called transversal. The eigenvectors may be arbitrarily chosen within the plane normal to $\vec{k}$. However, it is conventional to choose one vector in the vertical plane (denoted SV) and the other purely horizontal (denoted SH).

The solutions presented here are for elastic plane waves in an unbounded isotropic homogeneous medium. We obtain two types of plane waves (one P- and one S-wave) that may propagate in any direction specified by the corresponding wavenumber vector $\vec{k}$. In this case, P- and S-waves are uncoupled. We will observe a more complicated behavior once material interfaces or a free surface are introduced. At a free surface, the requirement is that the traction should vanish. This leads to a special wave that consists of coupled exponentially decaying P- and S-waves, the so-called Rayleigh wave. At an interface between two media, the boundary conditions are that

- the displacement $\vec{u}$ is continuous, and
- the traction $\vec{T}(\vec{n})$, associated with the normal $\vec{n}$ to the interface, is continuous.

These boundary conditions can only be satisfied if P- and S-waves are coupled at the interface, i. e., waves are not only scattered when impinging on a reflector, but also mode converted from P- into S-waves and vice versa. This, on the one hand, complicates the elastic wave propagation problem considerably but, on the other hand, scattered elastic wavefields contain a lot of information beyond scalar wavefields. However, working with elastic wavefields in migration and inversion has not been widely investigated and is topic of current research, see, e. g., Goertz (2002). Usually, a fluid medium is assumed in seismic prospecting for oil and gas as an approximation of the solid medium—this leads to the so-called acoustic case.

Applying the above-mentioned boundary conditions to a plane wave impinging on a plane reflector, one will derive the well-known Snell's law and the corresponding reflection and transmission coefficients. A detailed analysis is given in, e. g., Popov (2002).

In inhomogeneous media, there exist in general no independent P- and S-wave propagations, i. e., the compressional and the shear waves are coupled. This is due to the gradients of the Lamé parameters appearing in equation (2.10). Although the Earth is inhomogeneous rather than homogeneous, we can observe independent P- and S-wave arrivals in almost every seismic record. Actually, in slowly varying (so-called smooth) inhomogeneous media, we may approximately separate the wavefield into many independent elementary P- and S-wave contributions. In this case, we assume that the characteristic properties of the elastic medium remain almost constant in all directions within a wavelength interval (except, of course, at interfaces). Apparently, this approach is heuristical rather than strictly mathematical, but it allows us to understand the wave propagation phenomena more clearly.

If anisotropy is to be considered, things become more tedious. In an anisotropic homogeneous medium, three types of plane waves can propagate, namely one quasi-compressional (qP) wave and two quasi-shear (qS1 and qS2) waves with, in general, different properties, e. g., different propagation velocities. The observation of two independently traveling S-waves is usually called *shear wave splitting in anisotropic media.* For the isotropic case, two eigenvalues corresponding to the system of equations (2.13) are identical[1] which generally happens no longer if an anisotropic medium is considered. In general, the phase and group velocity vectors of any plane wave propagating in the anisotropic medium have different directions and magnitude, i. e., the phase velocity vector which is perpendicular to the wavefront, and the group velocity vector that is parallel to the energy flux, do not coincide. In terms of ray theory (see Chapter 3), we may say that rays are no longer orthogonal to the wavefronts. The velocities, of course, depend on the direction of propagation, and the behavior of the polarization vector differs significantly from the isotropic case. In addition, the phase velocity vectors of both qS1- and qS2-waves can coincide for certain specific propagation directions. Then, the behavior of qS-waves becomes particularly complex—such situations are called shear wave singularities.

[1] Such an occurrence is usually denoted as a degeneration.

2.5 The acoustic wave equation

A medium with $\mu = 0$ is called a fluid. If we neglect body forces, we immediately observe that in such media the elastodynamic wave equation (2.10) reduces to

$$\vec{\nabla}\left(\lambda\vec{\nabla}\cdot\vec{u}\right) = \rho\frac{\partial^2\vec{u}}{\partial t^2} \,. \tag{2.23}$$

However, rather than working with the displacement vector $\vec{u}(\vec{x},t)$, it is more common to work with the pressure P defined by

$$P(\vec{x},t) = -\lambda\vec{\nabla}\cdot\vec{u}(\vec{x},t) \,. \tag{2.24}$$

We then obtain from equation (2.23)

$$\vec{\nabla}\left(\frac{1}{\rho}\vec{\nabla}P\right) = \frac{1}{\lambda}\frac{\partial^2 P}{\partial t^2} \,, \tag{2.25}$$

which is known as the *acoustic wave equation*. For a medium with constant density ρ, the acoustic wave equation is given by

$$\frac{1}{v^2}\frac{\partial^2 P}{\partial t^2} - \Delta P = 0 \,, \quad v = \sqrt{\frac{\lambda}{\rho}} \,, \tag{2.26}$$

where v is the spatially varying acoustic wave velocity.

In the acoustic case, only one type of wave can propagate, namely the compressional wave. Shear waves do not propagate in a fluid medium.

2.6 Summary

In this chapter, we have seen how the general elastodynamic wave equation can be derived from Newton's second law of mechanics. Making some assumptions on the model, simplifications can be applied and we obtain, e. g., the acoustic wave equation which is frequently used in seismic prospecting for hydrocarbons as an approximation of the solid Earth. The formulas presented in this chapter are the governing equations for seismic reflection modeling as well as imaging and have, thus, a fundamental importance. In the next chapter, a high-frequency asymptotic method based on the wave equation will be presented that is frequently used for investigating both the forward and the inverse problems, namely the so-called ray method.

Chapter 3

Ray theory

There exist several methods to describe and to calculate the propagation of seismic (body) waves in complex inhomogeneous media. They can roughly be divided into

1. methods based on the direct solution of the wave equation, e. g., by means of numerical finite-difference (FD) methods, and

2. approximate high-frequency asymptotic methods, which leads us to ray theory.

The ray method is a powerful and frequently used tool for investigating both the forward and the inverse problems in reflection seismics, and comprehensive use of it is made throughout this thesis. The ray method has two main advantages: it provides a physical and intuitive insight into wave propagation phenomena and is rather computationally efficient compared to direct numerical solutions of the wave equation. However, it should also be mentioned that this method does not describe the whole wavefield and can suffer from some drawbacks which are closely related to focal regions, usually called caustics in geophysical terminology.

Some fundamental ideas of ray theory in the context of elastodynamics have been known in physics for a long time, and one of the first publications giving a consistent mathematical description might be Luneburg (1966). Since then, many papers and books have been published that cover important aspects and applications of the ray method. Among these are, e. g., Aki and Richards (1980), Hanyga (1984), Kravtsov and Orlov (1990), or Popov (2002). The most comprehensive description of the ray method can be found in Červený (2001). See also references therein.

Ray theory is an approximation and only valid if waves are of sufficiently high frequency. This statement sometimes confuses people because exploration geophysicists are usually dealing with frequencies in the range between 10 Hz and 100 Hz. One should note that the term "high frequency" does not refer to absolute values. What must be considered is the relationship between wavelengths (or wavenumbers, respectively) associated with the frequencies in the data, and the natural length scale (e. g., the thickness of layers, the radius of curvature of reflectors, or a measure of the inhomogeneity of material properties in the form $|\vec{\nabla} v|/v$, etc.) of the wave propagation medium. In order to be called high frequency, the length scales of interest in a medium should be many times as large as the dominant wavelength. However, the theory does not fail catastrophically if the high-frequency condition is not met. Failure of the theory will lead to incorrectly predicted amplitudes and growing errors in the location of reflectors or reflection events.

3.1 Eikonal and transport equations

For simplicity, we will now work with the acoustic wave equation (2.26). In the frequency domain, this equation reads

$$\left[\Delta + \frac{\omega^2}{v^2(\vec{x})}\right]\hat{P}(\vec{x},\omega) = 0\,, \tag{3.1}$$

where $\hat{P}(\vec{x},\omega)$ is the Fourier transformation of $P(\vec{x},t)$. Equation (3.1) is called *Helmholtz equation* and the operator defined by the squared brackets is known as the Helmholtz operator. Now, we make use of the WKBJ[1] trial solution (also known as the Debye series, or sometimes just called ray series in this context)

$$\hat{P}(\vec{x},\omega) \propto e^{i\omega\tau(\vec{x})}\sum_{j=0}^{\infty}\frac{A_j(\vec{x})}{(i\omega)^j}\,, \tag{3.2}$$

where $\tau(\vec{x})$ is called eikonal (actually, it is the traveltime), and the A_j's are the frequency-independent parameters representing the wave amplitude. These parameters need to be determined. Substituting the ansatz (3.2) in the Helmholtz equation gives

$$e^{i\omega\tau}\sum_{j=0}^{\infty}\frac{1}{(i\omega)^j}\left[\omega^2\left\{\frac{1}{v^2}-\left(\vec{\nabla}\tau\right)^2\right\}A_j + i\omega\left\{2\vec{\nabla}\tau\cdot\vec{\nabla}A_j + A_j\Delta\tau\right\} + \Delta A_j\right] = 0\,. \tag{3.3}$$

This equation must be satisfied for all frequencies. In general, as we cannot expect the terms of different powers of ω to cancel each other, the coefficients of the series must vanish independently. This leads to the *eikonal equation*

$$\left(\vec{\nabla}\tau(\vec{x})\right)^2 = \frac{1}{v^2(\vec{x})} \tag{3.4}$$

and the (first) *transport equation*

$$2\vec{\nabla}\tau(\vec{x})\cdot\vec{\nabla}A_0(\vec{x}) + A_0(\vec{x})\Delta\tau(\vec{x}) = 0\,. \tag{3.5}$$

After solving these two equations, one can construct the solutions to all other A_j's with $j > 0$. The term ΔA_j, however, remains in equation (3.3) and there is no chance to satisfy this equation exactly by ansatz (3.2). For high frequencies, we assume that ΔA_j is negligible compared to the other terms in equation (3.3) and so the wave equation is locally approximated by the eikonal and transport equations. If the high-frequency condition is not met or if the term ΔA_j increases along a ray, the approximation might break down. In this sense, we may understand this term as a criterion whether the ray method is a good approximation to the true wave propagation. For further details, see Červený (2001).

We cannot expect that the final solution yields a convergent series. Rather, we expect the result to be an asymptotic series for the exact solution with respect to ω tending to infinity, with only the first few terms actually being relevant. While for convergent series more terms give a better representation of a function, this statement does, in general, not hold for asymptotic series. The latter are often better representations than exact series results because there is no guarantee that a convergent representation will really converge rapidly. By design, asymptotic representations are forced to have initial terms that start out closer to the value of the function they represent. As a consequence, geophysicists usually

[1]The letters stand for the names Wentzel, Kramers, Brillouin, and Jeffreys, several of the many physicists who independently employed such representations.

stick to zero-order ray theory, i. e., only the amplitude term A_0 is considered. For further details about asymptotic series, see Bleistein (1984) or Bleistein and Handelsman (1986).

Up to now, all calculations were restricted to the acoustic case with the density ρ being constant. We can easily extend the theory to a variable-density medium. However, we will immediately observe that the variable density does not affect the kinematic part of our calculations, i. e., the eikonal equation remains unchanged. The amplitude behavior is, of course, affected but in such a way that we get the same type of differential equation as before. As a consequence, the amplitudes of the variable-density problem differ from the amplitude of the constant-density problem only by a factor that is proportional to $\sqrt{\rho(\vec{x})}$. Furthermore, the high-frequency elastic P- and S-wave propagation processes in a smooth inhomogeneous medium are also governed by the same types of eikonal and transport equations, although the calculations are more cumbersome. Therefore, there is no need to study all problems individually and for the purpose of this thesis I may restrict my considerations to the simplest case, namely the acoustic case.

3.2 Solution of the eikonal equation

Methods for solving the eikonal equation (3.4) can be found in any book dealing with partial differential equations, e. g., Bleistein (1984) or Bronstein and Semendjajew (1997). The most frequently used technique is the method of characteristics. In principle, we want to solve the problem

$$F(\vec{x}, \tau, \vec{p}) = 0 , \tag{3.6}$$

where F describes a seven-dimensional function depending on the space coordinates $\vec{x}$, the traveltime (or eikonal) τ, and the traveltime gradient (or slowness) $\vec{p} = \vec{\nabla}\tau$. This function, F, can be regarded as defining iso-surfaces in the (x_1, x_2, x_3)-space for each given constant value τ. These iso-surfaces are wavefronts with $\vec{p}$ being normal to them. Now, one can derive a system of ordinary differential equations that tells us how to move from one wavefront with $\tau = \tau_1$ to another one with $\tau = \tau_2$. The solutions for the spatial coordinates will then describe curves that are called *characteristics*, or, in our special case, *rays*. The solutions for the components of $\vec{p}$ tell us how that vector changes along each of the curves, while the solution for τ itself describes the change of τ along the characteristics.

Applying the method of characteristics to the eikonal equation (3.4) allows us to rewrite this partial differential equation as six ordinary differential equations, usually called *ray equations* or *kinematic ray tracing system*,

$$\begin{aligned} \frac{d\vec{x}}{d\tau} &= v^2 \vec{p} \, , \\ \frac{d\vec{p}}{d\tau} &= -\frac{\vec{\nabla} v}{v} \, , \end{aligned} \tag{3.7}$$

where v is the spatially varying velocity. We can also use the arclength s instead of the traveltime τ by substituting

$$\frac{d}{d\tau} = \frac{d}{ds}\frac{ds}{d\tau} = v\frac{d}{ds} \tag{3.8}$$

in equation (3.7).

The resulting linear ordinary differential equations (3.7) can now be integrated by a variety of methods depending on the problem and the desired degree of accuracy. It is interesting to note that we can solve

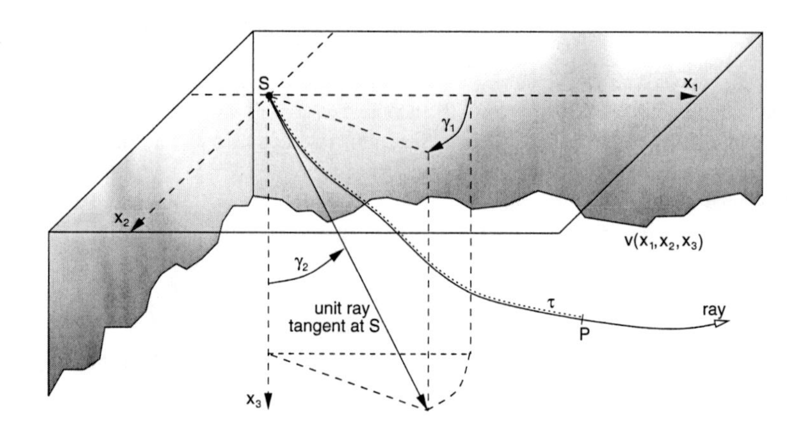

Figure 3.1: The ray coordinate system. For a point source S, ray coordinates in the region where the wavefield is regular are given by two angles, here denoted γ_1 and γ_2, and the eikonal, τ, that specifies the position of the point P on the ray. It is also possible to use the arclength s instead of the traveltime τ as the control variable γ_3 along the ray. Similar considerations lead to ray coordinates if an initial position on a wavefront is given.

for rays without ever determining the traveltime along the rays. To calculate the total traveltime along a ray, we need a seventh equation in addition to the six equations (3.7). Once a ray is determined, the traveltime is simply obtained by an integration of $ds/v(\vec{x})$ along the arbitrarily bent ray.

3.3 Solution of the transport equation

Solving the transport equation (3.5) is a little bit more cumbersome and can only be done when the eikonal equation has already been solved before. To simplify calculations, we introduce a new set of coordinates: the ray coordinates. These coordinates are given by two ray parameters γ_1 and γ_2, e. g., the angles of the spherical coordinate system, and one parameter γ_3 along the ray itself, e. g., the traveltime τ or the arclength s. For a fixed source point S and fixed angles γ_1 and γ_2, we depict one ray in the given wavefield, and the parameter $\gamma_3 = \tau$ (or s) indicates the position of a point P on this ray. Thus, each point P in the space Ω where the ray field is regular, can either be described by its Cartesian coordinates x_1, x_2, x_3 or the ray coordinates $\gamma_1, \gamma_2, \gamma_3$, see also Figure 3.1.

The components of the 3×3 transformation matrix $\hat{\mathbf{Q}}$ from ray coordinates γ_i to general Cartesian coordinates x_i are given by

$$\hat{Q}_{ij} = \frac{\partial x_i}{\partial \gamma_j} , \quad i, j = 1, 2, 3 , \tag{3.9}$$

where the actual form of the matrix depends on the choice of γ_3. The fundamental role in the investigation whether the ray field is regular is played by the Jacobian of the transformation, namely

$$J^{(\gamma)} = \left| \frac{\partial(x_1, x_2, x_3)}{\partial(\gamma_1, \gamma_2, \gamma_3)} \right| = \det\left(\hat{\mathbf{Q}}\right) , \tag{3.10}$$

where the superscript is used to indicate the control variable along the ray. If the Jacobian $J^{(\gamma)}$ is defined and does not vanish at any point of the space Ω, the ray field is called regular. Conversely, the ray field is called singular at any point where $J^{(\gamma)}$ is not defined or vanishes. If we consider the case where $\gamma_3 = s$, we call the corresponding Jacobian $J = J^{(s)}$ the *ray Jacobian*. In a geometrical interpretation, the ray Jacobian represents the element of volume dV in the new ray coordinate system and is, thus, a representation of the density of the rayfield.

Using the above introduced ray coordinates with $\gamma_3 = \tau$, we can rewrite the partial differential equation (3.5) as an ordinary differential equation along a ray,

$$\frac{2}{v^2}\frac{dA_0}{d\tau} + \frac{A_0}{vJ^{(\tau)}}\frac{d}{d\tau}\left(\frac{J^{(\tau)}}{v}\right) = 0 \,. \tag{3.11}$$

If $\gamma_3 = s$, equation (3.5) turns into

$$\frac{2}{v}\frac{dA_0}{ds} + \frac{A_0}{J}\frac{d}{ds}\left(\frac{J}{v}\right) = 0 \,. \tag{3.12}$$

Equation (3.12) or equation (3.11) can now easily be solved by separation of variables. Finally, we obtain the main term for the amplitude at a position defined by the arclength s along a ray,

$$A_0(s) = \frac{\Sigma(\gamma_1, \gamma_2)}{\sqrt{J(s)/v(s)}} = \sqrt{\frac{v(s)J(s_0)}{v(s_0)J(s)}}\,A_0(s_0) \,, \tag{3.13}$$

where Σ is the constant of integration depending on the ray take-off parameters only, and values with argument s_0 denote known (initial) data on a small sphere around a starting point, usually the source point. How to determine such initial ray data is discussed in, e. g., Červený (2001) or Popov (2002). We have, by now, reduced the solution of the transport equation (3.5) to the determination of the ray Jacobian J. Due to the fact that the rays and, thus, the density of the rayfield are assumed to be known, the calculation of J can be solved numerically.

The points of a ray where the ray Jacobian vanishes, i. e., $J = 0$, are called *caustic points* or *caustics*. A caustic point of first order is a point where one of the principal wavefront curvature radii becomes zero. In mathematical terms, $\mathrm{rank}(\hat{\mathbf{Q}}) = 2$ at caustic points of first order. Caustic points of second order are called focus points. Here, both principal wavefront curvature radii become zero and $\mathrm{rank}(\hat{\mathbf{Q}}) = 1$. At caustic points, a phase shift occurs that is dependent on the order of the caustic. In ray theory, the actually continuous transition of the wave phase across a caustic point is approximated by a discontinuous phase jump. The total phase shift due to caustics along a ray is given by $-\frac{1}{2}\pi\kappa$, where κ is the so-called KMAH index[2]. It increases by one for caustics of first order while caustics of second order count twice.

Note that the ray series ansatz is not valid in the vicinity of caustics. As J becomes zero, the amplitude A_0 would tend to infinity which makes no sense at all from a physical point of view. A method to overcome the caustic problem is known as the Gaussian beam method, see, e. g., Červený et al. (1982) or Popov (2002).

The quantity

$$\tilde{\mathcal{L}} = \sqrt{J} = \sqrt{|J|}\,e^{-i\frac{\pi}{2}\kappa} \tag{3.14}$$

is usually called the *geometrical spreading* factor in seismic ray theory. However, the terminology is not uniform in the literature. This factor plays an important role in true-amplitude imaging as was already pointed out in Chapter 1. Note that the choice of the sign in the exponent depends on the definition of the analytic signal (see, e. g., Taner et al., 1979).

[2]The KMAH index was introduced by Ziolkowski and Deschamps (1980), acknowledging the previous work of Keller, Maslov, Arnold, and Hörmander on this problem.

3.4 Paraxial and dynamic ray tracing

In the previous section, we have seen that the ray amplitude cannot be computed without the determination of the ray Jacobian J. In principle, J can be computed by analyzing the density of the rayfield. However, there exists another method to determine J and some other geometrical characteristics of the wavefront in the vicinity of a (known) ray by a procedure called dynamic ray tracing.

The dynamic ray-tracing system can be expressed in many forms and in various coordinate systems, whereas the simplest form is obtained in ray-centered coordinates. The ray-centered coordinate system q_1, q_2, and q_3 is a curvilinear orthogonal coordinate system introduced in such a way that the ray itself represents the third axis of the system. The other two axes q_1 and q_2 are formed by two mutually perpendicular straight lines intersecting at the ray and situated in a plane that is orthogonal to the ray at q_3. Thus, the plane q_3 = const is tangent to the wavefront, and the central ray itself is specified by the condition $q_1 = q_2 = 0$. Note that ray-centered coordinates are only suitable for describing points $\tilde{R}$ in the vicinity of a known ray as long as there exists only one plane perpendicular to the ray passing through $\tilde{R}$. This means that the region of validity of ray-centered coordinates depends on the curvature of the central ray. The region of validity is broad along slightly curved rays and narrow along rays with high curvatures. For more details on the ray-centered coordinate system and the computation of the basis vectors, see Popov and Pšenčík (1978).

In ray-centered coordinates, the eikonal equation (3.4) can be used to derive a simple approximate system of linear ordinary differential equations of the first order for rays that are situated in the vicinity of a known central ray. Such rays are called *paraxial rays* and the relevant system is called *paraxial ray-tracing system* according to the terminology in optics. The *dynamic ray-tracing system* is closely related to the paraxial system—in fact, the form of the system of differential equations is identical, only the calculated quantities have a different physical meaning. The term "dynamic ray-tracing system" was suggested by Červený and Hron (1980) as this system allows to calculate the geometrical spreading and curvatures of wavefronts which are important in evaluating the dynamic properties of seismic waves, i. e., ray amplitudes.

The paraxial ray-tracing system in ray-centered coordinates can be expressed as

$$\begin{aligned} \frac{dq_i}{ds} &= v p_i^{(q)} , \\ \frac{dp_i^{(q)}}{ds} &= -\frac{1}{v^2} q_j \left. \frac{\partial^2 v}{\partial q_i \partial q_j} \right|_{q_1=q_2=0} , \end{aligned} \tag{3.15}$$

where $i, j = 1, 2$ and $p_i^{(q)}$ represents the ray-centered covariant components of the slowness. It is a system of four linear differential equations. Note that the system of equations (3.15) is only an approximation and can only be used for rays that do not deviate considerably from the central ray. The dynamic ray-tracing system in its simplest form reads

$$\begin{aligned} \frac{d\mathbf{Q}}{ds} &= v\mathbf{P} , \\ \frac{d\mathbf{P}}{ds} &= -\frac{1}{v^2}\mathbf{V}\mathbf{Q} , \end{aligned} \tag{3.16}$$

with $\mathbf{V}$ being the matrix of second derivatives of the velocity v with respect to q_i along the central ray, i. e.,

$$V_{ij} = \left. \frac{\partial^2 v}{\partial q_i \partial q_j} \right|_{q_1=q_2=0} \qquad i, j = 1, 2 . \tag{3.17}$$

One immediately observes the close relationship between the systems of equations (3.15) and (3.16). The matrices **P** and **Q** are 2×2 transformation matrices with components

$$P_{ij} = \left. \frac{\partial p_i^{(q)}}{\partial \gamma_j} \right|_{q_1=q_2=0} , \qquad Q_{ij} = \left. \frac{\partial q_i}{\partial \gamma_j} \right|_{q_1=q_2=0} , \qquad i, j = 1, 2 . \tag{3.18}$$

The system (3.16) of ordinary linear differential equations of the first order has four linearly independent solutions. The fundamental matrix $\mathbf{\Pi}$ of the system that is formed of these four independent solutions is called the *propagator matrix of the dynamic ray-tracing system* or simply *ray propagator matrix*. Paraxial ray theory implies that the dynamic parameters of a point R on a paraxial ray are linearly dependent on those at a certain initial point S, i. e.,

$$\begin{pmatrix} q_1(R) \\ q_2(R) \\ p_1^{(q)}(R) \\ p_2^{(q)}(R) \end{pmatrix} = \mathbf{\Pi}(R, S) \begin{pmatrix} q_1(S) \\ q_2(S) \\ p_1^{(q)}(S) \\ p_2^{(q)}(S) \end{pmatrix} . \tag{3.19}$$

The matrix $\mathbf{\Pi}$ becomes the identity matrix at S on the central ray. Once the ray propagator matrix from S to R has been found along a ray, the solution of the dynamic ray-tracing system for any initial condition at S is obtained by merely multiplying the ray propagator matrix by the matrix of the initial conditions. Moreover, it is possible to obtain the geometrical spreading factor and other dynamic parameters of the wave propagation process from the ray propagator matrix. The propagator matrix is symplectic and can be chained along the central ray, see Červený (2001).

The 4×4 ray-centered propagator matrix $\mathbf{\Pi}$ of a ray connecting an initial point S (let us call it source point) and an end point R (which we call receiver point) can be expressed as

$$\mathbf{\Pi}(R, S) = \begin{pmatrix} \mathbf{Q_1} & \mathbf{Q_2} \\ \mathbf{P_1} & \mathbf{P_2} \end{pmatrix} . \tag{3.20}$$

The 2×2 matrices $\mathbf{Q_1}$, $\mathbf{Q_2}$, $\mathbf{P_1}$, and $\mathbf{P_2}$ can be obtained by solving the dynamic ray-tracing system (3.16) for special initial conditions. The matrices $\mathbf{Q_1}$ and $\mathbf{P_1}$ are obtained for the initial conditions

$$\mathbf{Q}(S) = \mathbf{I} , \quad \mathbf{P}(S) = \mathbf{0} , \tag{3.21}$$

where $\mathbf{I}$ and $\mathbf{0}$ are the 2×2 identity and zero matrix, respectively. This solution is also called the plane wave solution of the dynamic ray-tracing system. The matrices $\mathbf{Q_2}$ and $\mathbf{P_2}$ are obtained from the point source solution of the dynamic ray-tracing system with the initial conditions

$$\mathbf{Q}(S) = \mathbf{0} , \quad \mathbf{P}(S) = \mathbf{I} . \tag{3.22}$$

Note that the propagator matrix $\mathbf{\Pi}$ for fixed points S and R corresponding to the dynamic ray-tracing system (3.16) in ray-centered coordinates does not depend on the control variable along the ray, i. e., we can use either the traveltime τ or the arclength s without the need to modify the propagator matrix $\mathbf{\Pi}$.

The point source geometrical spreading factor can now be expressed as

$$\tilde{\mathcal{L}} = \sqrt{J} = \sqrt{\det(\mathbf{Q_2})} . \tag{3.23}$$

The *normalized geometrical spreading factor*[3] which is reciprocal is given by

$$\mathcal{L} = \frac{1}{\sqrt{v_S\, v_R}} \sqrt{\det(\mathbf{Q}_2)} \,, \tag{3.24}$$

where v_S and v_R are the wave propagation velocities at the source and receiver point, respectively. As already mentioned before, this factor plays an important role in the handling of dynamic effects in forward as well as inverse problems.

Sometimes, another alternative ray propagator matrix is used, the so-called surface-to-surface propagator matrix $\mathbf{T}$ which was introduced by Bortfeld (1989). The 4×4 propagator matrix $\mathbf{T}$ describes how the (projections of the) slowness vector $\vec{p}$ and the distance $\vec{x}$ between the paraxial and the central ray change as a result of wave propagation in the vicinity of the central ray starting at an anterior and ending at a posterior surface. The most important submatrix of $\mathbf{T}$ is the upper right 2×2 matrix $\mathbf{B}$ which corresponds to the submatrix $\mathbf{Q}_2$ of the ray-centered propagator matrix $\mathbf{\Pi}$. The matrix $\mathbf{B}$ relates to the geometrical spreading factor of a wave emanating from a point source. A complete description of the relationship between the surface-to-surface propagator matrix $\mathbf{T}$ and the ray-centered propagator matrix $\mathbf{\Pi}$ can be found in Hubral et al. (1992).

Using a notation related to the surface-to-surface propagator matrix $\mathbf{T}$, the normalized point source geometrical spreading factor can be written in the following form:

$$\mathcal{L} = \frac{\sqrt{\cos\alpha_S\, \cos\alpha_R}}{\sqrt{v_S\, v_R}} \sqrt{\det(\mathbf{B})} \,, \tag{3.25}$$

where α_S is the starting angle of the central ray at S and α_R is the emergence angle of the central ray at R (measured versus the normal of the anterior surface at S and the posterior surface at R, respectively). The parameters v_S and v_R are once again the wave propagation velocities at S and R, respectively (Schleicher, 1993).

In order to perform a true-amplitude migration (Chapter 5), we need to know a suitable approximation of the Green's function for the macro model with which to perform the migration process. A true-amplitude weight can be calculated from the dynamic part of the Green's function or, in other words, the geometrical spreading factor for any point P in the model and any fixed source location S must be known. Dynamic ray tracing is frequently used to determine such information. However, there are also ambitious efforts to use traveltime information only to evaluate the geometrical spreading factor and to calculate the true-amplitude weight function. Details about the underlying theory as well as some synthetic data examples can be found in, e. g., Vanelle (2002), see also references therein.

3.5 Summary

In this chapter, a high-frequency method called the ray method was presented to approximately describe the wave propagation process in smoothly varying inhomogeneous media. The fundamental equations that form the basis of the ray method are the eikonal and transport equations. By means of these equations, traveltimes and ray-theoretical amplitudes for seismic waves can be determined. Several extensions to standard ray tracing exist, e. g., paraxial or dynamic ray tracing. These methods are frequently used in the investigation of both the forward and the inverse problem in seismics. In the next chapter, we will have a closer look at the inverse problem in order to find an imaging procedure that allows us to transform the recorded seismic data into an image of the subsurface.

[3] Sometimes this factor is also called *relative geometrical spreading*.

Chapter 4

Inversion

In the introduction, it was pointed out that migration can be performed in various ways. Here, I focus on the Kirchhoff approach which is based on an integral solution of the wave equation. It was Schneider (1978) who related the diffraction-stack method of Hagedoorn (1954) to the wave equation and formulated a theory that is today known as *Kirchhoff migration*. Since then, many authors have provided valuable contributions to the theory and applicability of seismic inversion and (true-amplitude) Kirchhoff migration, see, e. g., Herman et al. (1986), Bleistein (1987), Berkhout (1987), Docherty (1991), or Schleicher et al. (1993) and references therein. For simplicity, only the acoustic wave equation will be considered here. However, the discussion applies in principle to the inversion of elastic or electromagnetic data as well. The following considerations mainly follow the lines of M. Popov (pers. comm.).

4.1 Introduction

An initial-boundary-value problem for the acoustic wave equation can be formulated as follows:

$$\left(\Delta - \frac{1}{v^2}\frac{\partial^2}{\partial t^2}\right) P\left(\vec{x}, t\right) = -f\left(\vec{x}, t\right) \;, \tag{4.1}$$

where the vector $\vec{x} = (x_1, x_2, x_3)$ lies within a space denoted by Ω and the time t varies between t_0 and T. Note that unlike in equation (2.26) the density of external forces is not omitted here. The initial conditions (or Cauchy data) are given at the time t_0 and may be written as

$$P\bigg|_{t=t_0} = P^{(0)}\left(\vec{x}\right) \;, \quad \frac{\partial}{\partial t} P\bigg|_{t=t_0} = P^{(1)}\left(\vec{x}\right) \;. \tag{4.2}$$

The boundary condition may either be the Dirichlet condition

$$P\bigg|_{\partial\Omega} = F_{(D)}\left(\vec{x}, t\right) \;, \quad \vec{x} \in \partial\Omega \;, \tag{4.3}$$

or the von Neumann condition

$$\frac{\partial P}{\partial n}\bigg|_{\partial\Omega} = \vec{\nabla} P \cdot \vec{n}\bigg|_{\partial\Omega} = F_{(N)}\left(\vec{x}, t\right) \;, \quad \vec{x} \in \partial\Omega \;, \tag{4.4}$$

where $\partial\Omega$ is the boundary surface of the space Ω, and $\vec{n}$ is the unit normal vector to $\partial\Omega$ pointing outwards. In addition, there is a third type of boundary condition that combines the Dirichlet and von Neumann conditions. However, we do not consider this type here. As mentioned in Chapter 1, for a direct problem the velocity $v(\vec{x})$ is given and we want to construct the wavefield $P(\vec{x}, t)$, $\vec{x} \in \Omega$ for a specific time t within the time interval $[t_0, T]$. The Green's function G for this problem is the solution of

$$\left(\Delta - \frac{1}{v^2}\frac{\partial^2}{\partial t^2}\right) G(\vec{x}, t \,|\, \vec{x}_0, t_0) = -\delta(t - t_0)\,\delta(\vec{x} - \vec{x}_0) \;, \quad \vec{x}, \vec{x}_0 \in \Omega \;, \tag{4.5}$$

$$G\Big|_{t<t_0} = 0 \;, \tag{4.6}$$

where, in addition, a boundary condition of the Dirichlet or von Neumann type must be considered. Note that the initial as well as the boundary conditions are homogeneous in case we calculate the Green's function. In this way, the Green's function describes the development of the wavefield propagating forward in time, generated by a point source at position $\vec{x}_0$ that was triggered at time t_0. It is worthwhile to note that the term "source" does not necessarily describe an actual seismic source in the field. The Green's function is reciprocal, i. e.,

$$G(\vec{x}, t \,|\, \vec{x}_0, t_0) = G(\vec{x}_0, -t_0 \,|\, \vec{x}, -t) \;. \tag{4.7}$$

4.2 Green's formula and Kirchhoff theorem

If any of the functions $f(\vec{x}, t)$, $P^{(0)}(\vec{x})$, or $P^{(1)}(\vec{x})$ change in equations (4.1) or (4.2), we have to solve the above-mentioned problem in principle from the beginning. Thus, an important question is whether we are able to obtain the solution for any functions f, $P^{(0)}$, and $P^{(1)}$ if we know a special solution to a similar problem with a specific point-source function instead of f. We can answer in the affirmative as long as we know the Green's function for the considered problem and make use of Green's formula. To derive the solution, we start off with equation (4.1) and assume that the wavefield P is causal which implies that the initial conditions are given at time $t = 0$:

$$P\Big|_{t=0} = P^{(0)}(\vec{x}) \;, \quad \frac{\partial}{\partial t}P\Big|_{t=0} = P^{(1)}(\vec{x}) \;. \tag{4.8}$$

For a representation theorem, we have to use a Green's function that looks backward in time, therefore we define the Green's function according to equation (4.5) but with the condition

$$G\Big|_{t>t_0} = 0 \;. \tag{4.9}$$

Now, we multiply (from the left-hand side) equation (4.1) by $G(\vec{x}, t \,|\, \vec{x}_0, t_0)$ and equation (4.5) by $P(\vec{x}, t)$ and perform a subtraction which leads us to

$$P\delta(t - t_0)\,\delta(\vec{x} - \vec{x}_0) = fG + G\Delta P - P\Delta G - \frac{1}{v^2}G\frac{\partial^2 P}{\partial t^2} + \frac{1}{v^2}P\frac{\partial^2 G}{\partial t^2} \;, \tag{4.10}$$

where the arguments of the functions have been dropped for simplicity. In order to eliminate the delta functions, we integrate this equation over time and space, where the space volume Ω is the obvious choice. Since we assume both P to be causal and G to satisfy condition (4.9), we obtain

$$\begin{aligned}
\int_0^{t'} dt \iiint_\Omega P(\vec{x},t)\,\delta(\vec{x}-\vec{x}_0)\,\delta(t-t_0)\,d\Omega &= \int_0^{t'} dt \iiint_\Omega f(\vec{x},t)\,G(\vec{x},t\,|\,\vec{x}_0,t_0)\,d\Omega \\
&+ \int_0^{t'} dt \iiint_\Omega \{G(\vec{x},t\,|\,\vec{x}_0,t_0)\,\Delta P(\vec{x},t) - P(\vec{x},t)\,\Delta G(\vec{x},t\,|\,\vec{x}_0,t_0)\}\,d\Omega \\
&- \int_0^{t'} dt \iiint_\Omega \frac{1}{v^2}\left\{G(\vec{x},t\,|\,\vec{x}_0,t_0)\,\frac{\partial^2 P(\vec{x},t)}{\partial t^2} - P(\vec{x},t)\,\frac{\partial^2 G(\vec{x},t\,|\,\vec{x}_0,t_0)}{\partial t^2}\right\} d\Omega\,,
\end{aligned} \tag{4.11}$$

where $t' = t_0 + \epsilon$ and we have to take the limit $\epsilon \to 0$. Now, let us take a closer look at the individual terms: the left-hand side of this equation obviously becomes $P(\vec{x}_0, t_0)$. The second term on the right-hand side may be reformulated by means of Green's second formula, i. e.,

$$\int_0^{t_0} dt \iiint_\Omega \{G\Delta P - P\Delta G\}\,d\Omega = \int_0^{t_0} dt \oiint_{\partial\Omega} \left\{G\frac{\partial P}{\partial n} - P\frac{\partial G}{\partial n}\right\} dS\,, \tag{4.12}$$

where $\vec{n}$ is the unit normal vector to the surface $\partial\Omega$ that points outwards. The third term on the right-hand side in equation (4.11) is

$$\begin{aligned}
- \iiint_\Omega \frac{d\Omega}{v^2} \int_0^{t'} \left\{G\frac{\partial^2 P}{\partial t^2} - P\frac{\partial^2 G}{\partial t^2}\right\} dt &= - \iiint_\Omega \frac{d\Omega}{v^2} \int_0^{t'} \frac{\partial}{\partial t}\left\{G\frac{\partial P}{\partial t} - P\frac{\partial G}{\partial t}\right\} dt = \\
= - \iiint_\Omega \frac{d\Omega}{v^2} \left\{G\frac{\partial P}{\partial t} - P\frac{\partial G}{\partial t}\right\}\bigg|_{t=t'} &+ \iiint_\Omega \frac{d\Omega}{v^2} \left\{G\frac{\partial P}{\partial t} - P\frac{\partial G}{\partial t}\right\}\bigg|_{t=0}\,.
\end{aligned} \tag{4.13}$$

In this equation, the first integral that is evaluated at $t = t'$ vanishes because $G(\vec{x},t'\,|\,\vec{x}_0,t_0) = G(\vec{x},t_0+\epsilon\,|\,\vec{x}_0,t_0) = 0$ due to equation (4.9), and the second integral becomes

$$\iiint_\Omega \frac{d\Omega}{v^2} \left\{G(\vec{x},0\,|\,\vec{x}_0,t_0)\,P^{(1)}(\vec{x}) - P^{(0)}(\vec{x})\,\frac{\partial G(\vec{x},0\,|\,\vec{x}_0,t_0)}{\partial t}\right\}. \tag{4.14}$$

Finally, by gathering all intermediate results we can simplify equation (4.11) and obtain

$$\begin{aligned}
P(\vec{x}_0,t_0) &= \int_0^{t_0} dt \iiint_\Omega f(\vec{x},t)\,G(\vec{x},t\,|\,\vec{x}_0,t_0)\,d\Omega \\
&+ \int_0^{t_0} dt \oiint_{\partial\Omega} \left\{G(\vec{x},t\,|\,\vec{x}_0,t_0)\,\frac{\partial P(\vec{x},t)}{\partial n} - P(\vec{x},t)\,\frac{\partial G(\vec{x},t\,|\,\vec{x}_0,t_0)}{\partial n}\right\} dS \\
&+ \iiint_\Omega \frac{1}{v^2}\left\{G(\vec{x},0\,|\,\vec{x}_0,t_0)\,P^{(1)}(\vec{x}) - P^{(0)}(\vec{x})\,\frac{\partial G(\vec{x},0\,|\,\vec{x}_0,t_0)}{\partial t}\right\} d\Omega\,.
\end{aligned} \tag{4.15}$$

This formula is usually known as the *Kirchhoff theorem.*

Let us now have a closer look at the Kirchhoff theorem, bearing in mind the actual field measurement. We shall assume that there are no sources in the space Ω. In addition, we may expect that the initial conditions at the moment $t = 0$ are homogeneous. In mathematical terminology, that means

$$f(\vec{x}, t) = 0 \quad \text{and} \quad P^{(0)}(\vec{x}) = P^{(1)}(\vec{x}) = 0 \, . \tag{4.16}$$

Thus, the Kirchhoff theorem (4.15) reduces to

$$P(\vec{x}_0, t_0) = \int_0^{t_0} dt \oiint_{\partial\Omega} \left\{ G(\vec{x}, t \,|\, \vec{x}_0, t_0) \frac{\partial P(\vec{x}, t)}{\partial n} - P(\vec{x}, t) \frac{\partial G(\vec{x}, t \,|\, \vec{x}_0, t_0)}{\partial n} \right\} dS \, . \tag{4.17}$$

Unfortunately, for the field experiment we cannot consider the space Ω to be a compact domain[1] and, therefore, only a part of the boundary $\partial\Omega$ is accessible which we denote by S. We have to ensure that contributions to integral (4.17) are eliminated for that part S_* of the closed boundary that is not at our disposal. In other words, we have to define S_* around the observation point $\vec{x}_0$ far enough for the corresponding virtual (unaccessible) Huygens secondary sources to have a negligible contribution to the wavefield at $\vec{x}_0$, see Figure 4.1. Moreover, $\partial P / \partial n$ is often not available[2] and, thus, the corresponding term has to be eliminated in equation (4.17). This can either be achieved by just neglecting the corresponding term or by imposing a Dirichlet boundary condition for the Green's function G, i. e., $G|_{\partial\Omega} = 0$. Though the latter immediately results in a much more sophisticated approximation for G. All modifications to the Kirchhoff theorem may be justified by more or less rigorous mathematical considerations. However, it will hardly be a reliable justification and we should remember that our equations are only approximations. The final formula for $G|_{\partial\Omega} = 0$ reads

$$P(\vec{x}_0, t_0) = -\int_0^{t_0} dt \iint_S P(\vec{x}, t) \frac{\partial G(\vec{x}, t \,|\, \vec{x}_0, t_0)}{\partial n} dS \, , \tag{4.18}$$

which is usually known as *Kirchhoff integral.* According to the above-mentioned considerations, the closed boundary $\partial\Omega$ has been replaced by a simple boundary surface denoted by S. This may be, e. g., a part of the Earth's surface. The Kirchhoff integral allows us to calculate the wavefield at an observation point $\vec{x}_0$ (within the space Ω) at time t_0 once the wavefield on the boundary and the Green's function are known.

Please note that many authors (e. g., Aki and Richards, 1980; Herman et al., 1986; Červený, 2001) exchange the coordinates $\vec{x}$ and $\vec{x}_0$ as well as t and t_0 in equation (4.18) in order to obtain a formula where the observation point and time are denoted by $(\vec{x}, t)$ instead of $(\vec{x}_0, t_0)$ which usually stand for the source coordinates. This step is performed to obtain a notation that is commonly used in the literature. However, this also implies that the wave equation has to be rewritten in coordinates $(\vec{x}_0, t_0)$. I will not follow this approach here.

[1] In mathematical terminology, a closed and bounded domain is called compact.

[2] It might be estimated in offshore measurements by means of two streamers that are towed in two different depths; however, an estimation of the term in onshore measurements is much more difficult.

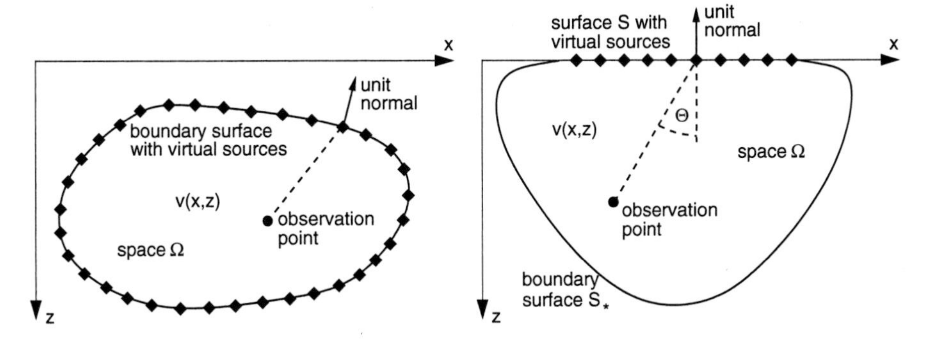

Figure 4.1: Principle of the Kirchhoff method (2D sketch). Left: the wavefield at any observation point $\vec{x}_0$ within the space Ω can be constructed as an integral over the boundary surface $\partial\Omega$ of the contributions of virtual sources placed along it. Right: the same situation depicted for seismic measurements. The boundary surface $\partial\Omega$ is split into two parts, the seismic surface S with virtual sources and an extension S_* around the observation point $\vec{x}_0$ far enough for the corresponding virtual (unaccessible) sources to have a negligible contribution to the wavefield at $\vec{x}_0$.

4.3 Reduction to the homogeneous case, direct problem

In this section, we assume that the velocity $v(\vec{x})$ is constant. We are then able to explicitely calculate the Green's function. If we take equation (4.9) into account, we obtain from equation (4.5)

$$G(\vec{x}, t \,|\, \vec{x}_0, t_0) = \frac{\delta\left(t - t_0 + \frac{r_0}{v}\right)}{4\pi r_0} \,, \tag{4.19}$$

where $r_0 = |\vec{x} - \vec{x}_0| = \sqrt{(x-x_0)^2 + (y-y_0)^2 + (z-z_0)^2}$. Let us now assume that the observation surface is given by $z = 0$. We are then able to derive the Green's function for the half-space that satisfies the Dirichlet boundary condition (i. e., $G|_{z=0} = 0$) by means of the mirror principle[3]. This Green's function is given by

$$G_{(D)}(\vec{x}, t \,|\, \vec{x}_0, t_0) = \frac{\delta\left(t - t_0 + \frac{r_0}{v}\right)}{4\pi r_0} - \frac{\delta\left(t - t_0 + \frac{r_*}{v}\right)}{4\pi r_*} \,, \tag{4.20}$$

where $r_* = \sqrt{(x-x_0)^2 + (y-y_0)^2 + (z+z_0)^2}$. For the problem under consideration, the Kirchhoff integral reads

$$P(\vec{x}_0, t_0) = -\int_0^{t_0} dt \iint_{(z=0)} P(\vec{x}, t) \frac{\partial G_{(D)}(\vec{x}, t \,|\, \vec{x}_0, t_0)}{\partial n} \, dS \,. \tag{4.21}$$

Obviously, $\partial/\partial n = -\partial/\partial z$ and

$$\frac{\partial}{\partial z}\delta\left(t - t_0 + \frac{r_0}{v}\right) = \frac{1}{v}\frac{\partial r_0}{\partial z}\delta'\left(t - t_0 + \frac{r_0}{v}\right) , \tag{4.22}$$

[3]Note that the mirror principle cannot be used if the seismic surface has topographic variations.

where the prime denotes a derivative with respect to the full argument. By means of the following equalities,

$$\left.\frac{\partial r_0}{\partial z}\right|_{z=0} = -\left.\frac{\partial r_*}{\partial z}\right|_{z=0} , \qquad \left. r_0\right|_{z=0} = \left. r_*\right|_{z=0} , \tag{4.23}$$

we obtain

$$\left.\frac{\partial G_{(D)}}{\partial n}\right|_{z=0} = -\left.\frac{\partial G_{(D)}}{\partial z}\right|_{z=0} \left.\left\{\frac{-1}{2\pi v r_0}\frac{\partial r_0}{\partial z}\delta'\left(t - t_0 + \frac{r_0}{v}\right) + \frac{1}{2\pi r_0^2}\frac{\partial r_0}{\partial z}\delta\left(t - t_0 + \frac{r_0}{v}\right)\right\}\right|_{z=0} . \tag{4.24}$$

Finally, we obtain from equation (4.21)

$$P(\vec{x}_0, t_0) = -\frac{1}{2\pi v}\iint\limits_{(z=0)} \frac{\partial r_0}{\partial z}\frac{P'\left(\vec{x}, t_0 - \frac{r_0}{v}\right)}{r_0}\, dS - \frac{1}{2\pi}\iint\limits_{(z=0)} \frac{\partial r_0}{\partial z}\frac{P\left(\vec{x}, t_0 - \frac{r_0}{v}\right)}{r_0^2}\, dS , \tag{4.25}$$

because

$$\int_0^{t_0} P(\vec{x}, t)\,\delta'\left(t - t_0 + \frac{r_0}{v}\right) dt = -\left.\frac{\partial P(\vec{x}, t)}{\partial t}\right|_{t=t_0-\frac{r_0}{v}} . \tag{4.26}$$

As long as we are concerned with deep reflectors, i. e., reflectors that have a certain distance from the measurement surface, we may neglect the second integral in equation (4.25) because it is of higher order in $1/r_0$. This leads us to the so-called *far-field approximation*. The term $\partial r_0/\partial z\,\big|_{z=0} = -\cos\Theta$ is known as *obliquity factor*. The parameter Θ denotes the angle between the vertical and the line connecting the (virtual) source point and the observation point, see Figure 4.1.

4.4 Inverse problem

For the inverse problem in seismics, we consider

$$\left(\Delta - \frac{1}{v^2}\frac{\partial^2}{\partial t^2}\right) P(\vec{x}, t) = 0 , \quad \vec{x} \in \Omega , \quad t \in [0, T] , \tag{4.27}$$

$$\left. P\right|_{t=T} = P^{(0)}(\vec{x}) , \quad \left.\frac{\partial P}{\partial t}\right|_{t=T} = P^{(1)}(\vec{x}) , \quad \vec{x} \in \Omega \tag{4.28}$$

$$\left. P\right|_{\partial\Omega} = P_{(D)}(\vec{x}, t) , \quad \vec{x} \in \partial\Omega , \tag{4.29}$$

where P denotes the wavefield generated by a seismic source, and T is the maximum measurement time. The velocity $v(\vec{x})$ is not known, except that it is supposed to be a smooth function in the space Ω. We shall now assume that $t \leq T$, i. e., we propagate the wavefield backward in time. For the derivation of the representation theorem, we thus need a Green's function that is defined according to equation (4.5) with the initial condition (4.6). Now, we may repeat the same steps as in the derivation of the representation theorem for the direct (forward) problem. If the initial data $P^{(0)}$ and $P^{(1)}$ are set to zero (which is usually done in seismic migration), we finally obtain

$$P(\vec{x}_0, t_0) = \int_{t_0}^{T} dt \oiint\limits_{\partial\Omega} \left\{G(\vec{x}, t \,|\, \vec{x}_0, t_0)\,\frac{\partial P(\vec{x}, t)}{\partial n} - P(\vec{x}, t)\,\frac{\partial G(\vec{x}, t \,|\, \vec{x}_0, t_0)}{\partial n}\right\} dS . \tag{4.30}$$

If we are able to construct the Green's function that satisfies the Dirichlet boundary condition $G|_{\partial\Omega} = 0$, equation (4.30) simplifies and reads

$$P(\vec{x}_0, t_0) = -\int_{t_0}^{T} dt \oiint_{\partial\Omega} P(\vec{x}, t) \frac{\partial G(\vec{x}, t \,|\, \vec{x}_0, t_0)}{\partial n} dS . \tag{4.31}$$

The integral is called *Porter-Bojarski integral* (Herman et al., 1986) and the equation describes the *Kirchhoff migration* process. We can once again impose some restrictions to the surface $\partial\Omega$ in order to replace the closed surface by a simple surface denoted by S (which is the measurement surface for migration purposes). Then, we have to ensure that no waves except those recorded on the measurement surface contribute to the migration result at the observation point $\vec{x}_0$, see also Figure 4.1. Please remember that the Green's function in equations (4.31) and (4.18) differ due to the different propagation direction.

4.5 Reduction to the homogeneous case, inverse problem

As in the case of the forward problem, we assume now that the boundary surface $\partial\Omega$ is just given by $z = 0$. The Green's function that satisfies the Dirichlet boundary condition reads

$$G_{(D)}(\vec{x}, t \,|\, \vec{x}_0, t_0) = \frac{\delta\left(t - t_0 - \frac{r_0}{v}\right)}{4\pi r_0} - \frac{\delta\left(t - t_0 - \frac{r_*}{v}\right)}{4\pi r_*} . \tag{4.32}$$

Note the inverted sign here compared to equation (4.20) which is due to the different initial conditions, i. e., this function now satisfies the condition $G = 0$ for $t < t_0$. This is the adjoint Green's function to the function given in equation (4.20). The derivative of this Green's function is given by

$$\left.\frac{\partial G_{(D)}}{\partial n}\right|_{z=0} = \left\{\frac{-1}{2\pi v r_0}\frac{\partial r_0}{\partial z}\delta'\left(t - t_0 - \frac{r_0}{v}\right) - \frac{1}{2\pi r_0^2}\frac{\partial r_0}{\partial z}\delta\left(t - t_0 - \frac{r_0}{v}\right)\right\}\Bigg|_{z=0} , \tag{4.33}$$

where I have used the corresponding derivative according to equation (4.22) and the equalities of equation (4.23). By inserting this expression into equation (4.31), we obtain

$$P(\vec{x}_0, t_0) = \frac{1}{2\pi v}\iint_{(z=0)} \frac{\partial r_0}{\partial z}\frac{P'\left(\vec{x}, t_0 + \frac{r_0}{v}\right)}{r_0} dS - \frac{1}{2\pi}\iint_{(z=0)} \frac{\partial r_0}{\partial z}\frac{P\left(\vec{x}, t_0 + \frac{r_0}{v}\right)}{r_0^2} dS . \tag{4.34}$$

Once again, it should be remarked that the second integral is often neglected if we are concerned with deep reflectors because of the higher order in $1/r_0$. Again, this simplification is known as far-field approximation in the literature. As previously stated, the term $\partial r_0/\partial z\big|_{z=0} = -\cos\Theta$ is called obliquity factor.

Equation (4.34) can be written in a different form which might be more familiar. By means of the equality

$$\left.\frac{\partial G_{(D)}}{\partial z}\right|_{z=0} = -\left.\frac{\partial G_{(D)}}{\partial z_0}\right|_{z=0} , \tag{4.35}$$

and using the far-field approximation, we are able to derive the following formula:

$$
\begin{aligned}
P(\vec{x}_0, t_0) &= \int_{t_0}^{T} dt \iint_{(z=0)} P(\vec{x}, t) \frac{-1}{2\pi} \frac{\partial}{\partial z_0} \frac{\delta\left(t - t_0 - \frac{r_0}{v}\right)}{r_0} dS \\
&= \frac{-1}{2\pi} \frac{\partial}{\partial z_0} \int_{t_0}^{T} dt \iint_{(z=0)} P(\vec{x}, t) \frac{\delta\left(t - t_0 - \frac{r_0}{v}\right)}{r_0} dS \qquad (4.36) \\
&= \frac{-1}{2\pi} \frac{\partial}{\partial z_0} \iint_{(z=0)} \frac{P\left(\vec{x}, t_0 + \frac{r_0}{v}\right)}{r_0} dS
\end{aligned}
$$

If we set $t_0 = 0$, the last equation is the 3D migration formula for stacked data (exploding reflector) of Schneider (1978).

4.6 General remarks

Let us now recall how the integrals that were presented in the previous sections allow us to calculate the wavefield at an observation point $\vec{x}_0$ from the wavefield known over time and over a given seismic surface. This is accomplished by summing the contribution of each virtual (Huygens secondary) source located within the seismic surface at the observation point $\vec{x}_0$, whereby the Green's function serves as a transfer function between the virtual source point and the observation point and correctly weights and delays the contributions. In order to obtain a migrated image (and not just a back-propagated wavefield), we need to apply a so-called *imaging condition*, see Claerbout (1971). While for a (hypothetical) exploding reflector experiment this imaging condition is fairly simple, it becomes more difficult for other types of experiments, see, e. g., Sullivan and Cohen (1987) or Docherty (1991) for details.

The main difference between the forward and the inverse problem is the Green's function which is (as stated in most books concerned with the topic of migration and inversion) either causal or anti-causal. Unfortunately, the terms causal and anti-causal depend on the way we observe the behavior of time. Usually, we observe a "movement forward in time". However, if we accept that a "forward movement back in time" may happen, we may call all Green's function causal because they are zero before waves appear. Note that this is different from a "backward movement in time". To clarify the situation, think of a person walking from a point A to a point B on a sidewalk. After having arrived at point B, the person wants to return to point A. Now, there are two possible solutions: a) the person may just go back without turning around (this corresponds to a backward propagation), or b) the person may turn around and go (forward) back to point A (this corresponds to a forward propagation back in time/space). From the point of view of the mathematical techniques involved here, the different initial conditions (4.6) and (4.9) for the Green's function provide the same result: they eliminate one of the terms in formula (4.13) which appeared after the integration over the time t. Let us now return to the geophysical terminology. The Kirchhoff integral is used to describe the (forward) propagation of seismic waves within a given depth model. However, the Kirchhoff integral itself cannot be used to solve the inverse problem, i. e., to describe backward propagation—this was already mentioned in the introduction. That is why Kirchhoff migration (the Porter-Bojarski integral) was introduced as adjoint operation that describes the forward propagation of the recorded wavefield in the reverse (time) direction.

In order to calculate the Green's function for migration, we need some knowledge about the velocities in the Earth prior to the migration process.[4] Such information is usually called a *macrovelocity model* as this model reflects only smooth large-scale features of the Earth. As mentioned in the introduction, we can extract such a model from the measured data, and we finally try to image the sharp discontinuities (reflectors) by a suitable imaging procedure called migration or inversion, respectively. Unfortunately, the Green's function can only be exactly calculated in homogeneous media (the results were presented in the previous sections). In inhomogeneous media, the spherical divergence effects can no longer be described by the term $1/r_0$ and the traveltime is no longer given by r_0/v. Instead of these simple terms, we shall use a ray-theoretical description based on the considerations in Chapter 3, which is given by

$$G(\vec{x}, t \,|\, \vec{x}_0, t_0) \quad \propto \quad A(\vec{x}, \vec{x}_0)\, \delta\left(t - t_0 - \tau(\vec{x}, \vec{x}_0)\right) , \tag{4.37}$$

where A is the Green's function amplitude which is proportional to the reciprocal geometrical spreading factor, $\mathcal{L}^{-1}(\vec{x}, \vec{x}_0)$, and τ is the traveltime (eikonal) from the source point to the observation point. In principle, there may be more than one ray connecting the source point and the observation point (multipathing). The complete ray-theoretical Green's function can be obtained by superposition of all elementary Green's functions corresponding to the different raypaths. This is important in the context of multiarrival Kirchhoff migration, see, e. g., Gray et al. (2001) and references therein.

4.7 Summary

In this chapter, integral solutions to the wave equation have been presented. The solution of the forward problem leads to the well-known Kirchhoff integral that is frequently used for modeling purposes. The solution of the inverse problem leads to the Porter-Bojarski integral and to a method called Kirchhoff migration. For the homogeneous case, the Green's function can be analytically calculated and, thus, the Porter-Bojarski integral can be simplified. In this way, the migration formula of Schneider (1978) was derived. In the next chapter, a more familiar representation of Kirchhoff migration is presented along with suitable weight functions to make the migrated image true-amplitude.

[4]This problem does not appear in a forward modeling process as the model and, thus, the velocities are given.

Chapter 5

True-amplitude Kirchhoff migration

In Chapter 4, a strict mathematical derivation of Kirchhoff migration was presented that is based on general inversion theory. However, there exists a geometrically more appealing way to describe Kirchhoff migration based on the concepts of surfaces of maximum convexity or surfaces of maximum concavity, respectively (Hagedoorn, 1954). These approaches are described in detail in this chapter along with the derivation of a weight function that makes the migration output true-amplitude. Thus, amplitudes in migrated images will be related to physical properties of the Earth.

5.1 Hagedoorn's imaging surfaces

I assume now that the seismic configuration is fixed, i. e., seismic traces are described according to equation (1.2) by the configuration vector $\vec{\xi}$ and selected configuration matrices $\mathbf{\Gamma}$. A point in the time domain, i. e., in the recorded seismograms, is, thus, denoted by $N(\vec{\xi}, t)$. Correspondingly, a point in the depth domain, i. e., in the image to be produced, is denoted by $M(\vec{r}, z)$, where $\vec{r} = (x, y)$.

Since the famous work of Hagedoorn (1954), concepts like the surface of maximum convexity which has become known as *diffraction traveltime surface*, *Huygens surface*, or *inplanat* as wells as concepts like the surface of maximum concavity that is also known as *isochron* or *aplanat* are well established in seismic imaging. The Huygens surface, on the one hand, is the kinematic image in the time domain of a point in the depth domain. In other words, for a fixed point $M(\vec{r}, z)$ in the depth domain and varying source-receiver pairs (S, R), the Huygens surface denotes the set of all points $N(\vec{\xi}, t = \tau_D)$ in the time domain for which t equals the sum of traveltimes from S to M and from M to R, i. e.,

$$\tau_D(\vec{\xi}, M) = \tau\left(S(\vec{\xi}), M\right) + \tau\left(M, R(\vec{\xi})\right) . \tag{5.1}$$

A physical interpretation of the above mentioned construction is that of a point diffractor at M which is illuminated by the seismic experiment under consideration, i. e., for a fixed configuration. The resulting traveltime surface in the time domain would exactly be the Huygens surface. The isochron, on the other hand, is the kinematic image in the depth domain of a point in the time domain. That means, the isochron for a fixed point $N(\vec{\xi}, t)$ is defined as the set of all points $M(\vec{r}, z = z_I)$ for which the sum of traveltimes τ_D, equation (5.1), equals the given time t. Therefore, the isochron is also called *surface of equal reflection time*. Note that both, the Huygens surface and the isochron, are defined by the very same traveltime function τ_D where different parameters are fixed. To obtain the Huygens surface, one has to fix the subsurface point M (i. e., the coordinates $\vec{r}$ and z) while for the construction of the isochron, the point N (i. e., the coordinates $\vec{\xi}$ and t) has to be fixed.

5.2 Hagedoorn's imaging condition

Hagedoorn (1954) showed that the following properties hold as long as the correct macrovelocity model is used for migration:

- The Huygens surface τ_D pertaining to an actual reflection point M_R and the reflection traveltime surface τ_R (for a primary reflection event) are tangent surfaces in the time domain.
- The isochron pertaining to a reflection event N_R and the reflector are tangent surfaces in the depth domain.

These tangency properties may be called *Hagedoorn's imaging conditions*[1], see also Figure 5.1. Later on, we will see how the tangency properties are treated in a mathematical way by means of the Method of Stationary Phase. Hagedoorn's imaging conditions are known as *dualities* in seismic imaging and were described in detail by Tygel et al. (1995).

5.3 Kirchhoff migration

The geometrically most appealing way to describe a Kirchhoff migration is the following: we think of the method as an operation that verifies whether a certain depth point M (an observation point at $\vec{x}_0$, if we use the notation of Chapter 4) acted as a reflection point M_R in the seismic experiment that took place, i. e., whether there is a primary reflection event in the data that corresponds to a reflection at M. For this purpose, one considers the subsurface region to be imaged (the so-called migration target zone) as being represented by an ensemble of points on a grid which is in general, but not necessarily, a regular one. We treat all such depth points M on the grid like a diffraction point. In the already estimated macrovelocity model, we can calculate the Green's function for each of these points M. From the kinematic part of the Green's function, we can then determine the configuration-specific Huygens surface. The amplitudes of the (filtered) input seismograms are summed-up (i. e., stacked) along the Huygens surface and assigned to the corresponding depth point M. This explains why the Kirchhoff migration scheme is also called a *diffraction stack*. As stated by Hagedoorn's imaging conditions, the reflection traveltime surface τ_R and the Huygens surface τ_D are tangent if and only if M is an actual reflection point M_R on the reflector. Consequently, one can expect that a diffraction stack will provide a significant result when $M = M_R$ due to constructive interference of the contributions. Otherwise, the stacking result will be negligible due to destructive interference. This well-known fact is proven by means of the Method of Stationary Phase.

If desired and as mentioned in the introduction, the effect of geometrical spreading can be removed from the migration output amplitudes by multiplying the data during the stack with a true-amplitude weight factor that is, in general, determined from the dynamic part of the Green's function. The weight function itself does not affect the kinematic part of the Kirchhoff migration process. The stacking

[1] In the seismic literature, the term "imaging condition" is traditionally used as defined in Claerbout (1971).

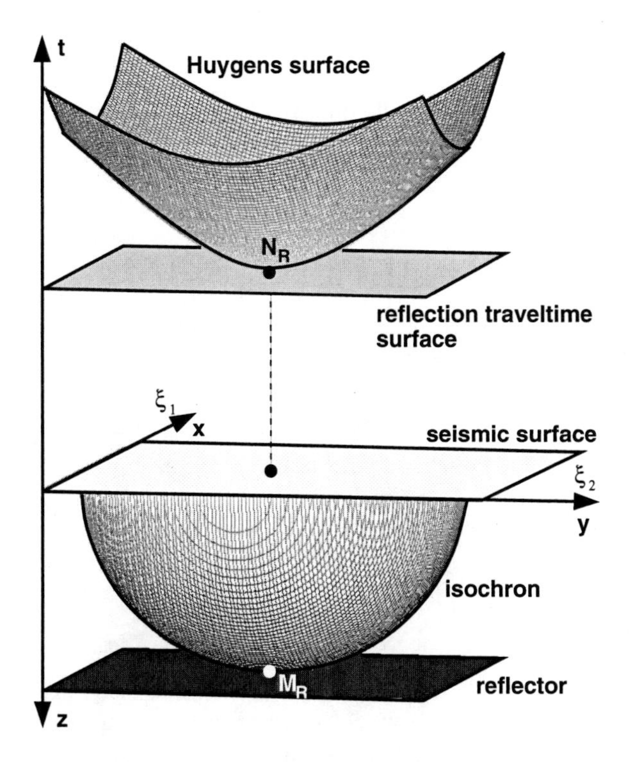

Figure 5.1: Dualities in migration, here indicated for the zero-offset configuration and a homogeneous overburden. Sketch of the reflection traveltime surface τ_R, the Huygens surface τ_D corresponding to the depth point M_R, the reflector, and the isochron corresponding to the point N_R in the time domain. The points N_R and M_R, the reflection traveltime surface and the reflector, and the Huygens surface and isochron, respectively, are called dual. Note that in the homogeneous case the Huygens surface is a hyperboloid in the time domain and the zero-offset isochron is a half-sphere in the depth domain.

surfaces are the same independent of whether we want to realize a simple structural migration or a complete true-amplitude migration with full weights applied. Note, however, that it is not sufficient to simply remove the geometrical spreading from the input traces firstly and subsequently apply an unweighted (kinematic) Kirchhoff migration. The reason is that the Kirchhoff migration process itself automatically introduces a partial geometrical-spreading correction that handles the effects due to the curvature of the reflector, called the *Fresnel geometrical-spreading factor* (Tygel et al., 1994a). Therefore, we need to find an appropriate weight function that accounts for the geometrical spreading along the ray segments SM and MR in the reflector's overburden, only. It will be independent of the reflector's properties and may, thus, be applied to any depth point M irrespective of whether it is an actual reflection point M_R.

Of course, the geometrical spreading is not the only factor affecting seismic amplitudes. For example, sources and receivers have certain characteristics that may vary with their position and influence the recorded amplitudes. Moreover, amplitudes are also expected to change due to transmission and attenuation in the overburden, see, e. g., Sheriff (1975). However, these additional amplitude effects do not pose any restrictions to the concept of true amplitudes. A true-amplitude imaging process will achieve its goal of correcting for geometrical spreading effects independently of other amplitude factors that may be present in the data. In practice, other factors than the geometrical spreading can often be neglected or they have been compensated for using some pre- or postprocessing methods. How, for instance, transmission losses can be successfully removed from seismic data has recently been shown by Hatchell (2000).

We assume now that the above-mentioned diffraction stack is the appropriate method to perform a true-amplitude migration and try to set up a formalism to derive the true-amplitude weight function. In principle, we can use any type of stacking integral as long as it correctly describes the migration process. Here, we use the following expression to describe the diffraction stack (Schleicher et al., 1993),

$$V(M) = \frac{-1}{2\pi} \iint_A d\xi_1 \, d\xi_2 \, W_{DS}(\vec{\xi}, M) \left. \frac{\partial U(\vec{\xi}, t)}{\partial t} \right|_{t=\tau_D(\vec{\xi},M)} . \tag{5.2}$$

The stacking surface $\tau_D(\vec{\xi}, M)$ is the Huygens surface defined in equation (5.1) with the configuration vector $\vec{\xi}$ varying over the aperture A which thereby provides the region of integration, see also Figure 1.6. From a mathematical point of view, the region of integration should ideally be the total ξ_1-ξ_2-plane. This is, of course, not possible because the aperture is always limited by the size of the acquisition area. In the presence of noise or taking aliasing into account, one should confine A to an even smaller region (the so-called migration aperture) than the maximum possible aperture defined by the seismic experiment. The resulting limited-aperture migration and effects related to the size of the migration aperture will be discussed in Chapter 6. The input $U(\vec{\xi}, t)$ is assumed to be describable according to equations (1.1) and (1.2). It was already mentioned that U is chosen complex in order to correctly handle phase shifts in migration. Therefore, the output value V of the integration process that is assigned to the depth point M is complex, too. For displaying migrated images, one generally utilizes only the real part of $V(M)$. The filter $\partial/\partial t$ (here, the filter function corresponds to a time derivative) is needed to correctly recover the source pulse (Newman, 1975). The function $W_{DS}(\vec{\xi}, M)$ is the true-amplitude weight, yet to be specified.

Note that the integral (5.2) is justified by nothing else than the fact that it describes the diffraction stack and it will solve our problem. However, the reader will immediately recognize the close relationship to the inversion formula (4.34) derived in Chapter 4 if the far-field approximation is applied. In this way, one can understand equation (5.2) as a generalization of the migration formula given by Schneider (1978).

Expression (5.2) is intended for the principal component of the particle displacement and, thus, is more general than a migration formula for pressure data. However, it only holds if the measurement surface is not a free surface. If no such data are available, the effect of the free surface has to be removed before applying a true-amplitude migration. How this can be done using conversion coefficients is discussed in Červený (2001).

Now, we consider the weight function $W_{DS}(\vec{\xi}, M)$ which is still undefined. The starting point for our derivation is a modified version of the diffraction stack integral (5.2). We introduce an artificial variable t that may vary. In other words, we consider the time-dependent stack

$$V(M,t) = \frac{-1}{2\pi} \iint_A d\xi_1\, d\xi_2\, W_{DS}(\vec{\xi}, M)\, \frac{\partial U(\vec{\xi}, t+\tau_D)}{\partial t} \tag{5.3}$$

for arbitrary values of t. Geometrically, the introduction of the artificial time variable t amounts to nothing more than to consider a continuous set of stacks that are carried out along stacking surfaces that are shifted by an amount t compared to the Huygens surface corresponding to M. In other words, for each point M we consider a time band within which consecutive stacks are performed. In this way, we can transform equation (5.3) into the frequency domain to apply a stationary-phase analysis. Of course, the actual (time-independent) migration result is obtained from the stack with no time shift, i. e., an imaging condition $t = 0$ is to be applied. The stacks along the shifted Huygens surfaces need never be carried out in practice, they are just introduced for the mathematical treatment.

If we expect the primary reflection event to be represented by equation (1.1), we can rewrite equation (5.3) in the form

$$V(M,t) = \frac{-1}{2\pi} \iint_A d\xi_1\, d\xi_2\, W_{DS}(\vec{\xi}, M)\, R_c \frac{\mathcal{A}}{\mathcal{L}}\, \frac{\partial F(t+\tau_{dif})}{\partial t}\,, \tag{5.4}$$

where $\tau_{dif}(\vec{\xi}, M)$ is the difference between the diffraction and reflection traveltime, i. e.,

$$\tau_{dif} = \tau_D - \tau_R\,. \tag{5.5}$$

Applying a Fourier transformation yields equation (5.4) in the frequency domain,

$$\hat{V}(M,\omega) = -\frac{i\omega}{2\pi}\, \hat{F}(\omega) \iint_A d\xi_1\, d\xi_2\, W_{DS}(\vec{\xi}, M)\, R_c \frac{\mathcal{A}}{\mathcal{L}}\, e^{i\omega\tau_{dif}}\,, \tag{5.6}$$

where $\hat{F}(\omega)$ and $\hat{V}(M,\omega)$ denote the Fourier transforms of $F(t)$ and $V(M,t)$, respectively. Remember the Fourier transformation theorems

$$\begin{aligned} F(t-\tau) \quad &\circ\!\!-\!\!\bullet \quad e^{-i\omega\tau}\, \hat{F}(\omega)\,, \\ \frac{d^n F(t)}{dt^n} \quad &\circ\!\!-\!\!\bullet \quad (i\omega)^n\, \hat{F}(\omega)\,, \end{aligned} \tag{5.7}$$

which were used here. Equation (5.6) cannot be solved analytically. It is, however, possible to evaluate the integrals by means of the 2D Method of Stationary Phase that provides an approximation to an integral of the form

$$I(\omega) = \iint_A d\xi_1\, d\xi_2\, f(\vec{\xi})\, e^{i\omega\tau(\vec{\xi})} \tag{5.8}$$

for sufficiently high frequencies ω. The restriction to high frequencies is in fact already implicitly done because all calculations are carried out within the framework of ray theory. Details about the 2D Method of Stationary Phase can be found in Bleistein (1984) or Bleistein and Handelsman (1986), a related explanation of the 1D Method of Stationary Phase can be found in Appendix A.

We assume that there exists one simple isolated stationary point $\vec{\xi}^*$ within the aperture A. Such a stationary point is defined according to

$$\vec{\nabla}\tau_{dif}\Big|_{\vec{\xi}=\vec{\xi}^*} = \vec{0}\,, \tag{5.9}$$

i. e., the gradient of the phase function τ_{dif} vanishes. Note that the gradient is taken with respect to $\vec{\xi}$. If no such stationary point exists within the aperture A, the main contributions to the integral stem from the boundary of the aperture. These unwanted contributions that might distort a migrated image are explained in detail in Chapter 6. It is possible to suppress these so-called aperture or boundary effects and, thus, the stack will yield a result that is zero or, at least, very close to zero in case no stationary point exists. Applying the stationary-phase method means to expand the phase function τ_{dif} of equation (5.6) into a Taylor series up to second order with respect to the stationary point $\vec{\xi}^*$, i. e.,

$$\tau_{dif}(\vec{\xi}, M) = \tau_{dif}(\vec{\xi}^*, M) + \frac{1}{2}(\vec{\xi} - \vec{\xi}^*)\mathbf{H}_{dif}(\vec{\xi} - \vec{\xi}^*)^T\,, \tag{5.10}$$

where the Hessian matrix $\mathbf{H}_{dif}$ is given by

$$\mathbf{H}_{dif} = \frac{\partial^2\tau_{dif}}{\partial\xi_i\partial\xi_j}\bigg|_{\vec{\xi}=\vec{\xi}^*}\,. \tag{5.11}$$

The superscript T denotes a transposed vector. In the following, we assume $\mathbf{H}_{dif}$ to be non-singular, i. e., $\det(\mathbf{H}_{dif})$ does not vanish. Then the analysis of equation (5.6) by means of the 2D Method of Stationary Phase (Bleistein, 1984) yields

$$\hat{V}(M,\omega) \approx \hat{F}(\omega)\; W_{DS}(\vec{\xi}^*, M)\; R_c\frac{\mathcal{A}}{\mathcal{L}}\; e^{i\omega\tau_{dif}}\; \frac{1}{\sqrt{|\det(\mathbf{H}_{dif})|}}\; e^{-i\frac{\pi}{2}(1-\frac{1}{2}\mathrm{Sgn}(\mathbf{H}_{dif}))}\,, \tag{5.12}$$

where $\mathrm{Sgn}(\mathbf{H}_{dif})$ is the signature of the matrix $\mathbf{H}_{dif}$, i. e., the number of positive eigenvalues minus the number of negative ones.

If we define

$$W_{DS}(\vec{\xi}^*, M) = \mathcal{L}\sqrt{|\det(\mathbf{H}_{dif})|}\; e^{i\frac{\pi}{2}(1-\frac{1}{2}\mathrm{Sgn}(\mathbf{H}_{dif}))}\,, \tag{5.13}$$

the migration output (5.12) reduces to

$$\hat{V}(M_R,\omega) \approx \hat{F}(\omega)\; R_c\mathcal{A}\; e^{i\omega\tau_{dif}}\,, \tag{5.14}$$

which is the spectrum of the source wavelet multiplied with the reflection coefficient and a phase-shift factor that accounts for the difference between the reflection and diffraction traveltime surface at the stationary point, see Figure 5.2.

Going back to the time domain and applying the imaging condition $t = 0$ yields

$$V(M) = V(M, t = 0) = R_c\mathcal{A}F(\tau_{dif})\,, \tag{5.15}$$

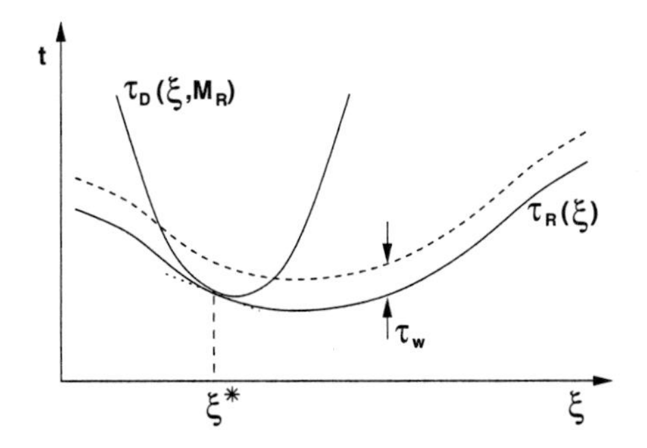

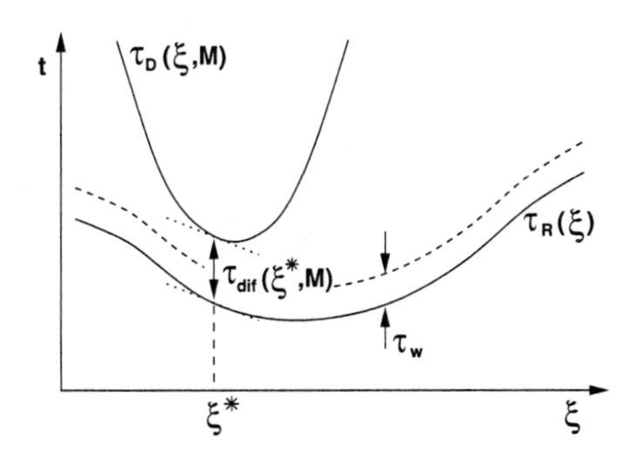

Figure 5.2: 2D sketch of possible stationary situations. The wavelet length in the time domain is given by τ_w. Left: the diffraction traveltime curve τ_D of an actual reflection point M_R is tangent to the reflection traveltime curve τ_R at the stationary point ξ^*. Right: the diffraction traveltime curve τ_D of a point M that lies deeper than the depth point M_R. Both the diffraction and reflection traveltime curves have the same slope at the stationary point ξ^*, but they are not tangent. The traveltime difference is given by τ_{dif}.

where $\tau_{dif} = \tau_{dif}(\vec{\xi}^*, M)$. This is the migration result where geometrical spreading effects have been removed if a stationary point exists within the aperture A; otherwise, the result will be zero. However, one should note that the source wavelet is a transient signal, i. e., $F(t)$ is zero outside an interval $0 \leq t \leq \tau_w$, where τ_w is the wavelet length. If $M = M_R$, then the diffraction and reflection traveltime surfaces are tangent at the stationary point, i. e., τ_{dif} vanishes and the diffraction stack provides the value $R_c\mathcal{A}F(0)$. Now, let us assume that M is a point dislocated in the vertical direction from the reflection point M_R. This corresponds to a small traveltime difference $t = \tau_{dif}$ with $0 \leq t \leq \tau_w$. The migration, thus, yields the value $R_c\mathcal{A}F(\tau_{dif})$. If the point M moves further away from the reflector, the result of the true-amplitude stack vanishes, see Figure 5.2. In this way, the complete wavelet is transferred from the time domain to the depth domain. It should, however, be emphasized, that a pulse distortion occurs when migrating seismic primary reflections obtained from arbitrary source-receiver configurations into depth, regardless of whether the macrovelocity model is correct or not. The relationship of the original time pulse and the depth pulse after migration can be expressed by a stretching factor

$$m_D = \left.\frac{\partial \tau_D}{\partial z}\right|_{M_R} = \frac{2}{v_{M_R}} \cos\alpha_{M_R} \cos\beta\,, \tag{5.16}$$

where v_{M_R} is the local velocity at the actual reflection point M_R, α_{M_R} is the reflection angle, and β is the reflector dip at M_R (Schleicher et al., 1993). The pulse distortion is the larger the smaller the stretch factor becomes as this factor appears in the argument of the pulse.

5.4 2.5D migration

If we consider 3D wave propagation in a medium that has only 2D parameter variations, i. e., the medium does not vary with respect to the coordinate perpendicular to the seismic line, we call it a 2.5D case. The direction of the acquisition line is usually denoted by ξ_1 in the time domain and by x in the depth domain and referred to as in-plane or inline direction whereas the ξ_2- and y-coordinate, respectively, defines the out-of-plane or crossline direction. As a consequence, shot and receiver positions depend only on the variable ξ_1 (the variable ξ_2 is fixed) and all possible reflection events stem from points within the vertical plane below the acquisition line, see also Figure 5.3. In the 2.5D case, we can simplify the migration formula (5.2) by making use of the medium's symmetry with respect to the seismic line located at the fixed ξ_2-position, see, e. g., Martins et al. (1997) or Bleistein et al. (2001) and references therein for a detailed explanation.

To derive the 2.5D diffraction stack integral, we recall the 3D formula (5.2) where the stack is carried out over the aperture A defined in the ξ_1-ξ_2-plane. As in the 2.5D case the data acquired in the ξ_1-direction do not depent on the location of the seismic line in the ξ_2-direction, the migration aperture A can be understood as an infinite strip in the crossline direction and the integration over ξ_2 in equation (5.2) can be solved analytically. Such simplification yields that the migration reduces to an in-plane stack, i. e., stacking surfaces shrink to stacking curves. We can rewrite equation (5.2) as

$$V(M) = \frac{1}{2\pi} \int_{a_1}^{a_2} d\xi_1 \; I_{DS}(\xi_1, \xi_2, M) \, , \qquad a_1 \leq \xi_1 \leq a_2 \tag{5.17}$$

where I_{DS} is given by

$$I_{DS}(\xi_1, \xi_2, M) = - \int_{-\infty}^{+\infty} d\xi_2 \; W_{DS}(\xi_1, \xi_2, M) \left. \frac{\partial U(\xi_1, \xi_2, t)}{\partial t} \right|_{t=\tau_D(\xi_1, \xi_2, M)} . \tag{5.18}$$

Evaluating the integral I_{DS} by means of the Method of Stationary Phase yields

$$I_{DS}(\xi_1, \xi_2, M) \approx \sqrt{2\pi} \left(\left. \frac{\partial^2 \tau_D}{\partial \xi_2^2} \right|_{(\xi_1, 0, M)} \right)^{-\frac{1}{2}} \left. \partial_t^{\frac{1}{2}} U(\xi_1, 0, t) \right|_{t=\tau_D(\xi_1, 0, M)} . \tag{5.19}$$

The filter function that was given by a time derivative in the 3D case reduces to a so-called half-derivative[2] denoted by $\partial_t^{1/2}$ which equals a multiplication with $\sqrt{-i\omega}$ in the frequency domain. An additional factor $\left(\partial^2 \tau_D / \partial \xi_2^2\right)^{-1/2}$ occurs which has to be applied to the weight function. Finally, we obtain the following integral for the 2.5 D case:

$$V(M) = \frac{1}{\sqrt{2\pi}} \int_{a_1}^{a_2} d\xi \; W_{DS}^{2.5D}(\xi, M) \left. \partial_t^{\frac{1}{2}} U(\xi, t) \right|_{t=\tau_D(\xi, M)} , \tag{5.20}$$

where $(\xi_1, 0)$ was replaced by ξ for simplicity. The region of integration is now given by a_1 and a_2 which corresponds to the aperture in the 2.5D case. The 2.5D weight function can be obtained from the 3D weight function by

$$W^{2.5D} = W^{3D} \left(\frac{1}{\sigma_S} + \frac{1}{\sigma_G} \right)^{-\frac{1}{2}} , \tag{5.21}$$

where σ_S and σ_R denote the out-of-plane spreading factors for the ray segments SM and MR, respectively. The out-of-plane spreading factor is given by the integral

$$\sigma = \int_{\text{ray}} v(s) \, ds \tag{5.22}$$

[2] Actually, it is the Hilbert transform of a half-derivative, see Bleistein et al. (2001) for details. However, this fact is usually ignored in most literature.

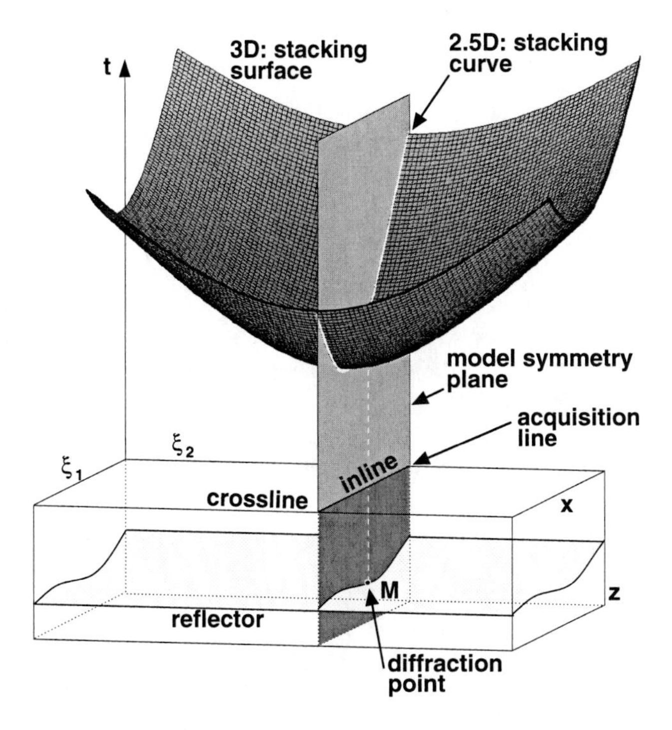

Figure 5.3: The difference between 3D and 2.5D diffraction stacks, here indicated for the zero-offset configuration and a homogeneous reflector overburden. In principle, to obtain the migration output value for a depth point M one has to stack amplitudes of the filtered input data along the corresponding complete Huygens surface τ_D. Making use of the fact that subsurface features do not change in crossline direction (2.5D situation), is is sufficient to stack along the Huygens curve. The stacking surface (3D), thus, shrinks to a stacking curve (2.5D) because the stack in the crossline direction can be evaluated analytically by means of the stationary-phase method and is taken into account by the out-of-plane geometrical spreading factor.

with s being the arclength and v the wave propagation velocity along the ray. In addition, the equalities

$$\left.\frac{\partial \tau_D}{\partial \xi_2}\right|_{\xi_2=0} = 0 \qquad \text{and} \qquad \left.\frac{\partial^2 \tau_D}{\partial \xi_2^2}\right|_{\xi_2=0} = \frac{1}{\sigma_S} + \frac{1}{\sigma_G} \tag{5.23}$$

were used to express $\partial^2 \tau_D / \partial \xi_2^2$ in terms of the out-of-plane spreading factors.

5.5 Comments on the weight function

It is to be shown in the following that the weight function (5.13) does not depend on reflector properties. Let us assume that M is an actual reflection point, i. e., $M = M_R$. The geometrical spreading factor can be decomposed according to Hubral et al. (1995) as

$$\mathcal{L}(S,R) = \frac{\mathcal{L}(S,M_R)\,\mathcal{L}(M_R,R)}{\mathcal{L}_F} = \frac{\mathcal{L}_{SM_R}\,\mathcal{L}_{M_RR}}{\mathcal{L}_F}\,, \tag{5.24}$$

where $\mathcal{L}_{SM_R}$ and $\mathcal{L}_{M_RR}$ are the point-source geometrical spreading factors of the two ray branches SM_R and M_RR, respectively. The factor $\mathcal{L}_F$ is the previously mentioned Fresnel factor and accounts for the influence of the Fresnel zone at M_R on the total geometrical spreading factor. The Fresnel factor can be written in terms of the matrix $\mathbf{H}_{dif}$ and reads

$$\mathcal{L}_F = \frac{m_D \cos\alpha_{M_R}}{h_B v_{M_R} \cos\beta} \sqrt{\left|\det(\mathbf{H}_{dif})\right|}\; e^{i\frac{\pi}{2}\left(1-\frac{1}{2}\mathrm{Sgn}(\mathbf{H}_{dif})\right)}\,. \tag{5.25}$$

The parameter $h_B = h_B(\vec{\xi}^*, M_R)$ is the Beylkin determinant (Beylkin, 1985a,b) given by

$$h_B(\vec{\xi}^*, M_R) = \det \left|\begin{matrix} \vec{\nabla}\tau_D(\vec{\xi}^*, M_R) \\ \frac{\partial}{\partial \xi_1}\vec{\nabla}\tau_D(\vec{\xi}, M_R) \\ \frac{\partial}{\partial \xi_2}\vec{\nabla}\tau_D(\vec{\xi}, M_R) \end{matrix}\right|_{\vec{\xi}=\vec{\xi}^*}\,, \tag{5.26}$$

where the gradient is taken with respect to depth coordinates at M_R. The parameter β is the reflector dip angle and α_{M_R} is the reflection angle at M_R. The velocity at M_R is given by v_{M_R}. If we insert the geometrical spreading decomposition (5.24) and the definition of the Fresnel factor (5.25) into equation (5.13), we obtain

$$W_{DS}(\vec{\xi}^*, M_R) = \frac{\mathcal{L}_{SM_R}\,\mathcal{L}_{M_RR}\, h_B v_{M_R} \cos\beta}{m_D \cos\alpha_{M_R}}\,. \tag{5.27}$$

Now we can use equation (5.16) to replace the vertical stretch factor in equation (5.27) which leads us to

$$W_{DS}(\vec{\xi}^*, M_R) = \frac{h_B\, v_{M_R}^2}{2\cos^2\alpha_{M_R}}\,\mathcal{L}_{SM_R}\,\mathcal{L}_{M_RR}\,. \tag{5.28}$$

This equation does only depend on the two individual ray branches SM_R and M_RR and not on any reflector properties. Now let us see how we can generalize the results. According to the fact that the weight function is independent of any reflector properties, we can use equation (5.28) not only for points M_R but for any arbitrary depth point M. We define β as the angle between the vertical and the half-angle direction between the two ray segments SM and MR. Furthermore, the geometrical spreading factor decomposition (5.24) is also valid irrespective of whether the point M is an actual reflection point or not. As a consequence, we are able to generalize equation (5.28) and we obtain the final result for the true-amplitude weight function, namely[3]

$$W_{DS}(\vec{\xi}, M) = \frac{h_B \, v_M^2}{2 \cos^2 \alpha_M} \, \mathcal{L}_{SM} \, \mathcal{L}_{MR} \,, \tag{5.29}$$

where $h_B = h_B(\vec{\xi}, M)$ is defined according to equation (5.26). It is important to note that the weight function on its own does not fully remove the geometrical spreading effect. Only the true-amplitude weight function *and* the summation process along the diffraction surface account for the total geometrical spreading effect. Surprisingly, the evaluation of the integral (5.2) by means of the stationary-phase method shows that the stack automatically corrects for the Fresnel factor, i. e., reflector curvature effects on the total geometrical spreading factor are automatically taken into account. The chosen weight function (5.29) simply corrects for the remaining geometrical spreading effects in the reflector's overburden. This is the reason why the weight function is independent of the reflector's curvature.

There exists various ways to express the true-amplitude weight function. All of them are, in principle, equivalent and only the notation differs depending on the physical properties of the wave propagation process that are used for the description. A number of expressions for different measurement configurations can be found in Schleicher et al. (1993) and Tygel et al. (1996). Note that the weight function for the zero-offset (ZO) configuration is remarkably simple,

$$W_{DS}^{ZO} = 4 \frac{\cos \alpha_S}{v_S} (-1)^{\kappa} \,, \tag{5.30}$$

where α_S is the ray-take off angle at the source location (which is also the receiver location in case of zero offset), v_S represents the associated wave propagation velocity, and κ denotes the KMAH index. Hanitzsch (1997) as well as Sun and Gajewski (1997, 1998) compare weight functions from different authors for a variety of shot-receiver configurations and a similar independent comparison was presented by Gray (1997). Martins et al. (1997) deal with weight functions for the 2.5D case and show simplifications for media with a constant velocity and a constant velocity gradient. Vanelle and Gajewski (2002a,b) show how the true-amplitude weight function can be determined from second order traveltime derivatives.

5.6 Alternative migration procedure

Kirchhoff true-amplitude migration cannot only be performed by a weighted stack along Huygens surfaces but also by smear-stacking along isochrons (Hubral et al., 1996). To understand this process think about points N in the time domain ($\vec{\xi}$-t-volume) distributed on a regular grid. Then construct for all these points the corresponding isochrons and distribute (i. e., smear out) the amplitude of the filtered seismic trace found at N along the related isochron. This means the amplitude value at N is assigned (with certain weights to account for the geometrical spreading effect) to each depth point M on the isochron. Afterwards, for each depth point M all assigned contributions are summed up—that is why such a procedure is called smear-stack. In this way, the approach relies on the constructive and destructive interference of all the isochrons in the image space. Some image points M end up

[3]The formula (11) in Tygel et al. (1996) is wrong, see also Jaramillo et al. (1998).

with large amplitude values as migration result and, thus, describe the location of reflectors within the migration target zone while other image points have negligible amplitudes due to the destructive interference of all contributions. This migration procedure is completely equivalent to the diffraction-stack migration described above (Hubral et al., 1996) and closely related to the previously mentioned dualities in seismic migration (Tygel et al., 1995).

5.7 Summary

In this chapter, a geometrically appealing way to perform a Kirchhoff migration process based on the concept of Huygens surfaces was presented. In addition, a weight function was derived that is applied in the diffraction stack and makes the output true-amplitude, i. e., amplitudes in migrated images are free of geometrical spreading effects. The practical procedure of 3D true-amplitude depth migration can be summarized as follows (Schleicher, 1993):

1. Filter each trace and transform the trace into a complex one, usually called the analytic signal, by means of the Hilbert transform. The filter function for 3D migration is given by a time derivative.

2. Determine the (migration) aperture A which consists of a 2D grid of points in the ξ_1-ξ_2-plane.

3. In the subsurface migration target zone, distribute points M forming a 3D grid. For simplicity, this grid should be spaced regularly.

4. Compute the traveltimes and dynamic parameters (such as the geometrical spreading factors) from all surface points S and R to all subsurface points M. This corresponds to the determination of the Green's function.

5. Repeat the following steps for every subsurface point M in the target zone. The result is a 3D depth-migrated section.

    (a) Determine from the kinematic part of the previously calculated Green's functions the diffraction traveltime τ_D for all $\vec{\xi}$ in A as the sum of traveltimes along the ray branches SM and RM where S and R are specified by $\vec{\xi}$.

    (b) Determine from the dynamic part of the previously calculated Green's functions the true-amplitude weight function $W_{DS}(\vec{\xi}, M)$ for all $\vec{\xi}$ in A. If, for one particular ray, a caustic appears at a surface point S or R, the weight of this ray is undefined and cannot be used.

    (c) Multiply the amplitude of the filtered analytic traces at time τ_D with $W_{DS}(\vec{\xi}, M)$ and sum them up for all $\vec{\xi}$ in A.

    (d) Assign the resulting stack signal to the corresponding point M. If the point M lies upon a reflector, the stack provides a value proportional to the complex angle-dependent reflection coefficient R_c times the amplitude factor $\mathcal{A}$.

In the next chapter, aperture effects are discussed in a mathematical way and related to the geometrical properties of Kirchhoff migration. In addition, a way is presented to suppress the boundary effects that might distort a migrated image.

Chapter 6

Migration aperture effects

Seismic images obtained by Kirchhoff migration are always accompanied by some artifacts known as "migration noise", "migration boundary effects", or "diffraction smiles" which may severely affect the quality of the migration result. Most of these undesirable effects are caused by a limited aperture, if the algorithms make no special disposition to avoid them. Likewise, a strong amplitude variation along reflection events may also cause similar artifacts. All these effects can be explained mathematically by means of the Method of Stationary Phase. However, such a purely theoretical explication is not always easy to understand for applied geophysicists. In this chapter, a geometrical interpretation of the terms of the stationary-phase approximation in relation to the diffraction and reflection traveltime surfaces in the time domain is presented.

6.1 Theoretical background

On the basis of the mathematical treatment of Kirchhoff migration (Schneider, 1978; Bleistein, 1987; Schleicher et al., 1993, see also Chapters 4 and 5), it became possible to study the boundary effects by means of the Method of Stationary Phase (see, e.g., Bleistein, 1984; Sun, 1998; Bleistein et al., 2001). However, these analytical studies have far less geometrical appeal than Hagedoorn's original insights. Therefore, it is desirable to connect the analytical aspects of Kirchhoff migration to its geometrical properties in order to provide a better understanding of the boundary effects.

As mentioned in Chapter 5, the extent of the Huygens surfaces, that is, the migration aperture, should be limitless so that no contributions due to the abrupt truncation of the integration or sum, respectively, occur. In practice, of course, the aperture is always limited by the region over which seismic data have been acquired. In other words, because of the finiteness of the survey area, Kirchhoff migration will always be a limited-aperture migration (LAM) (Sun, 1998). However, this is not the only reason why we have to deal with the effects of a finite migration aperture. In practical migration implementations, it is often advantageous to further restrict the aperture, i. e., to exclude certain source and receiver positions from the computations where data actually have been acquired. The main reasons for such an aperture restriction are

1. the use of fewer traces in the migration summation accelerates the entire migration process,
2. a smaller operator excludes steeper operator dips, which helps to avoid operator aliasing (see, e. g., Abma et al., 1999), and

3. less summation of data away from the signal reduces the stacking of unwanted noise.

For the best possible reduction of aliasing and noise as well as the best computational efficiency, one would like to use a model-based aperture restriction, i. e., the projected Fresnel zone. This zone is obtained by a projection of the actual Fresnel zone at the reflector along the rayfield onto the measurement surface (for details, see Hubral et al., 1993). The projected Fresnel zone coincides with the minimal migration aperture that correctly recovers true amplitudes (Schleicher et al., 1997; Sun, 2000; Sun and Bancroft, 2001). Unfortunately, it is difficult to determine the exact location and size of the Fresnel zone for each depth point prior to or during migration. A reasonable compromise between accuracy and practicability is to fix a migration aperture or a maximum migration dip. Note that a fixed aperture will image smaller maximum dips with increasing depth. On the other hand, to migrate a fixed maximum dip at all depths—like, e. g., in a 45° migration—requires an increasing migration aperture with depth. In regions where dips are known to be restricted, these convenient ways reduce aliasing and improve computational efficiency at the same time. However, close to the maximum dip, these dip-restricted migration operators will achieve only kinematically correct images (see, e. g., Schleicher et al., 1997; Sun, 1998). For true-amplitude migration, the maximum operator dip must always be chosen somewhat larger than the maximum reflector dip to be imaged.

The fact that the migration aperture is limited causes artifacts known as migration noise, boundary or aperture effects, or migration smiles. A *smile* is

> a concave-upward, semicircular event in seismic data that has the appearance of a smile and can be be caused by poor data migration or migration of noise.[1]

In this chapter, I relate the mathematical explanation of the migration artifacts by means of the Method of Stationary Phase (see, e. g., Felsen and Marcuvitz, 1973; James, 1976; Bleistein, 1984; Sun, 1998; Bleistein et al., 2001) to simple geometrical situations. In this way, my aim is to provide a more intuitive insight into these effects. Of course, since the stacking operations are the same in Kirchhoff time and depth migration, the corresponding artifacts are conceptually identical in both processes. Thus, I dedicate my present discussion only to Kirchhoff depth migration. It should, however, be kept in mind that everything said and shown in this chapter with respect to an image in depth holds in the same way for an image in time.

For simplicity, I restrict the following analysis to the 2.5D case. Conceptually, there is no difference in the application of the Method of Stationary Phase to the 3D migration integrals. The qualitative discussion involves the same arguments and leads to the same conclusions. The quantitative analysis is similar but slightly more complicated, mainly resulting in a different amplitude behavior of the artifacts.

As we have seen in Chapter 5, the 2.5D Kirchhoff migration process is mathematically expressed as an integration over the recorded wavefield. Here, I write down equation (5.20) once again,

$$V(M) = \frac{1}{\sqrt{2\pi}} \int_A d\xi \; W_{DS}^{2.5D}(\xi, M) \, \partial_t^{\frac{1}{2}} U(\xi, t)\Big|_{t=\tau_D(\xi,M)} ,$$

to allow a better understanding of the following derivation. All parameters were previously explained.

[1] Source: oilfield glossary, http://www.glossary.oilfield.slb.com/

I assume that at least one reflection event is present in the seismic data $U(\xi, t)$ that can be described according to equation (1.1). A seismic trace with several (primary) events may be described by superposition of individual seismic events. Upon substitution of expression (1.1), a time-dependent version of equation (5.20) reads in the frequency domain

$$\hat{V}(M,\omega) = \sqrt{\frac{-i\omega}{2\pi}}\,\hat{F}(\omega) \int_{a_1}^{a_2} d\xi\; W_{DS}^{(2.5)}(\xi, M)\, R_c \frac{\mathcal{A}}{\mathcal{L}}\, e^{i\omega\tau_{dif}}\,, \tag{6.1}$$

where $\tau_{dif}(\xi, M)$ again denotes the difference between the diffraction and reflection traveltimes, i. e., $\tau_{dif} = \tau_D - \tau_R$. The integral limits a_1 and a_2 denote the boundaries of the migration aperture. The migration result $V(M)$ is obtained from $\hat{V}(M,\omega)$ by an inverse Fourier transform together with the imaging condition $t = 0$.

In general, the integral in equation (6.1) cannot be solved analytically. The Method of Stationary Phase provides a way of analyzing the main contributions. Although in principle a high-frequency approximation, the Method of Stationary Phase yields highly accurate predictions of the migration results in the seismic frequency range. As mentioned before, the mathematical prerequisites for applying the Method of Stationary Phase are fulfilled implicitly since we perform all calculations within the framework of zero-order ray theory which is strictly valid only for high frequencies. Reducing it to its basic structure, the integral in equation (6.1) can be written in the form

$$I(\omega) = \int_{a_1}^{a_2} f(\xi) e^{i\omega q(\xi)} d\xi\,. \tag{6.2}$$

The Method of Stationary Phase is based on the observation that for high frequencies, i. e., for large values of ω, the factor $e^{i\omega q(\xi)}$ oscillates very rapidly, thus covering full periods in very small intervals of ξ. If $f(\xi)$ is not itself an oscillating function, its values do not strongly vary in any such interval. Thus, the integration over a full period of $e^{i\omega q(\xi)}$ yields approximately zero and does not contribute to the overall value of the integral. The only regions where $e^{i\omega q(\xi)}$ does not oscillate are those where the phase function $q(\xi)$ remains approximately constant or *stationary*. Mathematically, points of stationary phase are those where the phase function $q(\xi)$ has a horizontal tangent, i. e., a vanishing derivative. Nonnegligible contributions to integral (6.2) are, therefore, to be expected from the vicinity of these points. Further contributions to this integral are to be expected from the boundaries of the integration interval because there, the integration generally does not cover a full period of $e^{i\omega q(\xi)}$.

To illustrate these observations, I consider the migration of zero-offset data for a simple Earth model with a horizontal reflector at a depth of 1 km. For a point $M = M_R$ at $x = 3$ km on the reflector and a frequency of 30 Hz, Figure 6.1(a) shows the phase q of the integrand in equation (6.1) as a function of ξ. The dashed line in Figure 6.1(b) depicts the corresponding amplitude function $f(\xi)$. The solid line in Figure 6.1(b) shows the real part of the full integrand function. Note that this function strongly oscillates everywhere except in the vicinity of the point where the phase is stationary. It is evident that the amplitude modulation does not alter the oscillatory character of the integrand function.

Let us now discuss integral (6.2) in a more quantitative way. In our case, the phase function q is the difference between the diffraction and reflection traveltime curves, τ_{dif}. Thus, the real part of the integrand function (Figure 6.1(b)) has zeroes at

$$|\tau_{dif}| = |\tau_D - \tau_R| = n\frac{\pi}{\omega} = n\frac{T}{2}\,, \tag{6.3}$$

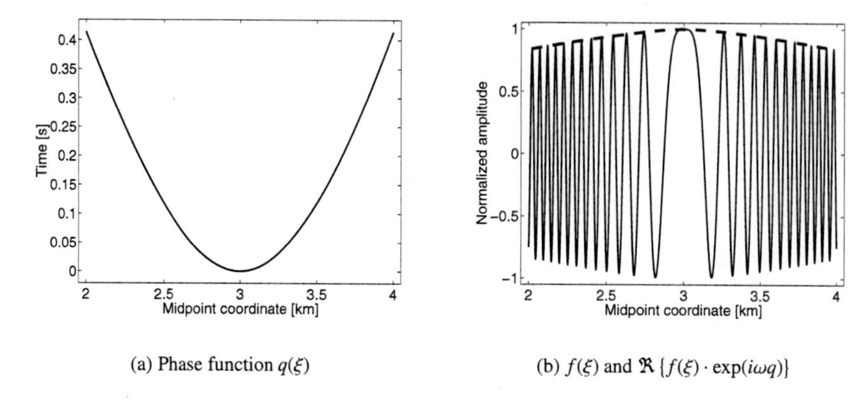

(a) Phase function $q(\xi)$

(b) $f(\xi)$ and $\Re\{f(\xi) \cdot \exp(i\omega q)\}$

Figure 6.1: Illustration of the integrand in equation (6.2). (a) Phase function $q(\xi)$. (b) Amplitude function $f(\xi)$ (dashed line) and real part of the complete integrand function $f(\xi) \cdot \exp(i\omega q)$ (solid line).

where $T = 2\pi/\omega$ is the period of the monofrequency wave under consideration. Equation (6.3) is equivalent to the definition of the boundary of the nth Fresnel zone (see, e.g., Červený and Soares, 1992). Therefore, the alternating zones of negative and positive amplitude of the integrand function are physically equivalent to the Fresnel zones. To be exact, the Fresnel zone is measured along the reflector, while the integration in equation (6.2) is carried out along the Huygens curve in the time domain. To see its influence on the integration, the Fresnel zone must be projected to the Earth's surface (Hubral et al., 1993). The true Fresnel zone in depth can be observed in a fully analogous manner in the Kirchhoff-Helmholtz modeling integral.

Now, consider an integration of the function $f \cdot \exp(i\omega q)$ from the center (where $\tau_D = \tau_R$) to the sides. At first, this sums up positive contributions from the first Fresnel zone, ending at the first zero in either direction. Subsequent Fresnel zones, each ending at the next zero, will add purely negative or positive contributions to integral (6.2). In other words, Fresnel zones with odd numbers contribute positively to the integral while Fresnel zones with even numbers contribute negatively. Because of the above observation that an integration over a full period, i. e., over two consecutive Fresnel zones, yields approximately zero, it becomes clear why the principal contribution to integral (6.2) will stem from the vicinity of the stationary point. Hence, an integration over only the first Fresnel zone already provides a very good approximation of the total integral. On the other hand, its full value cannot be recovered, if the integration interval does not completely cover the first Fresnel zone. This observation has lead to the quantification of the minimum migration aperture (Schleicher et al., 1997).

It has to be noted, however, that the above discussion holds strictly only for a monofrequency signal. For a transient, band-limited signal, one has to replace the half-period $T/2$ in equation (6.3) by some estimate τ_w of the wavelet length.

An analysis of the migration integral (6.2) by means of the Method of Stationary Phase under the assumption of a single, simple, and isolated point of stationary phase was carried out, e.g., by Bleistein (1984) or Sun (1998), see also Appendix A for details. Provided that f and q' are continuous functions of ξ, the analysis shows that the leading-order contributions to $I(\omega)$ are

$$I(\omega) \approx f(\xi^*)e^{i\omega q(\xi^*)}\sqrt{\frac{2\pi}{-i\omega q''(\xi^*)}} + \frac{1}{i\omega}\left[\frac{f(a_2)}{q'(a_2)}e^{i\omega q(a_2)} - \frac{f(a_1)}{q'(a_1)}e^{i\omega q(a_1)}\right], \quad (6.4)$$

where the prime denotes the derivative with respect to ξ. The point of stationary phase, defined by the condition $q'(\xi) = 0$, is denoted as ξ^*. We see that the terms in equation (6.4) are of the order $1/\sqrt{\omega}$ and $1/\omega$, respectively. For high frequencies, the two terms in equation (6.4) describe the major contributions to the final migrated image.

The first term of equation (6.4), which stems from the stationary point ξ^*, is of lower order in $1/\sqrt{\omega}$ than the second term. It will thus generally represent the dominant part of the total migrated section wherever it is nonzero. This term constitutes the actual migrated image of the reflector.

The second term of equation (6.4) is proportional to $1/\omega$. It is related to the endpoints of the integration/stacking operator. It is this second contribution that describes the main migration artifacts. Because of the higher order in $1/\sqrt{\omega}$, its amplitudes generally will be lower than those of the reflector image. However, under certain circumstances these effects can be as strong as (or even stronger than) a reflector image. This can happen when the amplitudes of the unmigrated section at the data margins (these enter into $f(a_1)$ or $f(a_2)$) are significantly larger than those at the stationary points (which enter into $f(\xi^*)$). The situation is the more probable the lower the frequencies contained in the unmigrated data.

Equation (6.4) must be slightly modified if the stationary point coincides with one of the boundaries, i. e., if $q'(a_1) = 0$ or $q'(a_2) = 0$. The corresponding boundary contribution at a_1 or a_2 is eliminated and the leading term is divided by 2. Note that the decay of the amplitudes across the aperture boundary is not abrupt but in the sense of an error function (Felsen and Marcuvitz, 1973).

If there are N simple, isolated stationary points of the phase function $q(\xi)$ in the integration interval (a_1, a_2), the interval is divided into N parts where each one contains exactly one stationary point. The independent analysis of each of the separate integrals can then be carried out as before, yielding the sum of all individual stationary-point contributions. The artificially introduced integral boundaries do not contribute to the final value of $I(\omega)$, because the corresponding boundary terms cancel each other.

Apart from the edges of the acquisition aperture and the stacking operator, also abrupt amplitude or phase changes along the reflection events in the seismic data may cause the same kind of endpoint contributions. These changes cause discontinuities in the amplitude or phase functions of integral (6.2) which make the integral act piecewise on the data. In other words, we may say that artificial endpoints are created which cause the additional aperture effects.

To describe these effects mathematically by means of the Method of Stationary Phase, I suppose such discontinuities to happen at J points d_j ($j = 1, ..., J$) within the integration interval (a_1, a_2). Then, the integration interval has to be further divided at all points d_j. The contributions from these new integration boundaries do not cancel because of the different values of f, q, and/or q' on either sides of the discontinuities. Thus, one obtains the general expression

$$\begin{aligned} I(\omega) \approx\ & \sum_{n=1}^{N} f(\xi_n^*)e^{i\omega q(\xi_n^*)}\sqrt{\frac{2\pi}{-i\omega q''(\xi_n^*)}} + \frac{1}{i\omega}\left[\frac{f(a_2)}{q'(a_2)}e^{i\omega q(a_2)} - \frac{f(a_1)}{q'(a_1)}e^{i\omega q(a_1)}\right] \\ & + \frac{1}{i\omega}\sum_{j=1}^{J}\left[\frac{f(d_j^-)}{q'(d_j^-)}e^{i\omega q(d_j^-)} - \frac{f(d_j^+)}{q'(d_j^+)}e^{i\omega q(d_j^+)}\right]. \end{aligned} \quad (6.5)$$

Here, $f(d_j^{\mp})$ denotes the values of f at the right- and left-hand sides of the discontinuity, i. e.,

$$f(d_j^{\mp}) = \lim_{\xi \to d_j^{\pm}} f(\xi) , \tag{6.6}$$

with corresponding definitions for $q(d_j^{\mp})$ and $q'(d_j^{\mp})$. For expression (6.5) to be valid, all stationary points ξ_n^* and all discontinuity points d_j have to be isolated from each other.

There are several situations in which such discontinuities can occur. They can be caused, e. g., by illumination problems or missing traces in the data. In this case, the migration smiles can even be desirable as they may help to recover reflector continuity. Moreover, amplitude variations along the reflection event which may be caused by focusing and defocusing effects of the reflected wave or attenuation acting differently on adjacent parts of the reflection event, can also cause similar effects as endpoints.

Migration artifacts due to a limited aperture, illumination problems, attenuation, or missing traces are inherent to seismic migration, independent of the actual migration scheme used. However, artifacts due to strong amplitude variations because of focusing effects are a consequence of Kirchhoff migration and can be largely reduced with other migration schemes such as, e. g., finite-difference wave-equation migration. In addition, Kirchhoff algorithms based on ray-tracing methods do not refract waves around obstacles; hence, these methods are more sensitive to illumination problems.

In contrast to the data boundaries, reflectors actually terminating in the Earth do not provoke migration smiles. In this case, edge diffractions are present in the seismic data that are collapsed by migration into the endpoint of the reflector. Because of the diffractions, the reflection event in the data has no actual endpoint but dies off over a larger number of traces. In this way, endpoint contributions are suppressed. The latter observation already points toward a well-known way of suppressing migration artifacts: tapering. I discuss this in a later section.

For the geometrical interpretation of the migration artifacts, we need the final migration result in depth. This is obtained by a multiplication of equation (6.4) with the factors in front of the integral in equation (6.1), together with an inverse Fourier transform under consideration of the imaging condition $t = 0$. The result is

$$V(M) \approx \frac{f(\xi^*)}{\sqrt{q''(\xi^*)}} F(q(\xi^*)) - \frac{1}{\sqrt{2\pi}} \frac{f(a_2)}{q'(a_2)} F^{\frac{1}{2}}(q(a_2)) + \frac{1}{\sqrt{2\pi}} \frac{f(a_1)}{q'(a_1)} F^{\frac{1}{2}}(q(a_1)) \tag{6.7}$$

with

$$f = W_{DS}^{(2.5)} R_c \frac{\mathcal{A}}{\mathcal{L}} \qquad \text{and} \qquad q = \tau_{dif} \; . \tag{6.8}$$

Moreover, $F^{\frac{1}{2}}(t)$ is a wavelet such that $\partial_t^{\frac{1}{2}} F^{\frac{1}{2}}(t) = F(t)$.

The first term of equation (6.7) stems from the stationary point ξ^* of the phase function $q = \tau_{dif} = \tau_D - \tau_R$, that is, the point where the difference between the diffraction and reflection traveltimes has an extremum. Thus, if the modulus of this traveltime difference $\tau_{dif}(\xi^*)$ is smaller than half the length τ_w of the wavelet $F(t)$, this contribution will be nonzero and, hence, constitute the actual migrated image of the reflector which, in general, represents the dominant part of the total migrated section. On the other hand, at points ξ^* where $|\tau_{dif}(\xi^*)| > \tau_w/2$, this term will be zero. Correspondingly, the boundary terms will vanish if $|\tau_{dif}(a_1)| > \tau_w/2$ and $|\tau_{dif}(a_2)| > \tau_w/2$, respectively[2]. I use this observation in the next section where I geometrically interpret formula (6.7).

[2]Here, I have assumed $F(t)$ to represent a zero-phase wavelet which differs from zero only in the interval $(-\tau_w/2, \tau_w/2)$. A corresponding formulation holds for a causal wavelet using the interval $(0, \tau_w)$.

6.2 Geometrical explanation of the aperture effects

Migration aperture effects are most easily explained by means of a simple numerical experiment for poststack data. The model consists of two half-spaces separated by a horizontal interface. The velocities in the upper and lower half-spaces are $v_p^{(1)} = 2$ km/s and $v_p^{(2)} = 3$ km/s, respectively, and the shear wave velocities are given by $v_s = v_p/\sqrt{3}$. The density is constant in the whole model. The zero-offset section was generated by dynamic ray tracing using a zero-phase Ricker wavelet with a dominant frequency of 20 Hz, a time sampling interval of $\Delta t = 1$ ms and a trace distance of $\Delta\xi = 5$ m. It was migrated with a 2.5D Kirchhoff true-amplitude depth migration scheme on a dense target grid ($\Delta x = 10$ m, $\Delta z = 2$ m) using the true velocity model.

For this simple model, the stacking operator is given by a hyperbola. I limit its spatial extent to a radius of 800 m with respect to the horizontal coordinate of the apex. In this way, the number of traces contributing to the stack for each depth point is 320. The migration target zone is placed at the end of the survey line to show the boundary effects. No effort is made to enhance or reduce the migration artifacts. The resulting migrated image is depicted in Figure 6.2. The red and blue colors in Figure 6.2 represent the positive and negative parts, respectively, of the signal as indicated in the zoomed area on the right side of the Figure.

By means of Figure 6.2, I am now going to discuss the boundary effects from a geometrical point of view, which allows us to gain a more intuitive insight. I then relate them to the above discussion of the interference in integral (6.2) and to the result of its stationary-phase evaluation as given by equation (6.7). For this purpose, I discuss the position of the Huygens curves pertaining to a series of characteristic depth points M_1 to M_9.

6.2.1 Points on the reflector: M_1

The actual reflector (which is unknown prior to migration) is built up by depth points such as M_1. The pertinent Huygens curves are tangent to the reflection traveltime curve[3]. Thus, amplitudes gathered along such curves sum up coherently and provide high stacking results that are assigned to the corresponding depth points. No boundary effects are present because the input data at the endpoints of the stacking operator, which correspond to the limits of integration a_1 and a_2 in equation (6.1), are zero. Of course, in practice there will always be some endpoint contributions because of the noise inherent in the seismograms.

I now relate this physical explanation to my earlier considerations of the Method of Stationary Phase. The stationary-phase condition $q'(\xi^*) = 0$ is satisfied where τ_D and τ_R have equal dip. Thus, we identify the horizontal coordinates of the tangency points with the points of stationary phase ξ^*. The value assigned to M_1 is mathematically described by the first term in equation (6.7). The second and third (boundary) terms of the stationary-phase result (6.7) vanish. The reason is that the differences between the diffraction and reflection traveltimes at $\xi = a_1$ and $\xi = a_2$ are larger than half the wavelet length, which implies that both $F^{\frac{1}{2}}(q(a_1))$ and $F^{\frac{1}{2}}(q(a_2))$ are zero.

[3]Note that in general, for laterally inhomogeneous media, the tangency points do not coincide with the apices of the stacking curves.

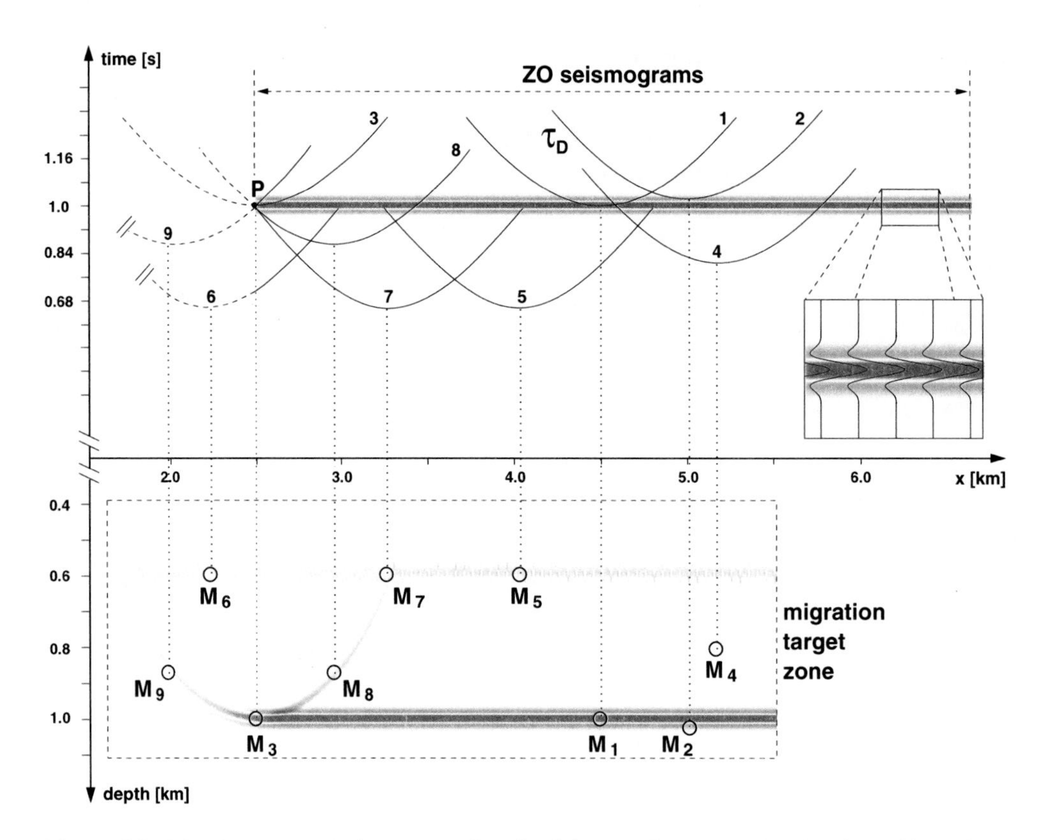

Figure 6.2: ZO seismograms and corresponding depth image after poststack migration with a migration aperture limited to 1600 m. Several characteristic depth points M_j and their pertinent stacking operators are shown. These are used to give a simple geometrical explanation of the limited-aperture migration effects. The zoomed area on the right side shows the zero-phase Ricker wavelet used in the modeling.

6.2.2 Points very close to the reflector: M_2

Points like M_2 also comprise part of the migrated image of the reflector. At the point where the diffraction traveltime curve of point M_2 has the same dip as the reflection traveltime curve, the former falls within a wavelet's length from the latter. Thus, migration acts almost as at M_1, i. e., amplitudes gathered along the stacking operator constructively interfere, in this way participating in the reconstruction of the source wavelet at the reflector image. The ratio between the vertical distance of point M_2 to the reflector and the shortest distance between the diffraction and reflection traveltime curves is the migration stretch factor m_D (Tygel et al., 1994b), already mentioned in Chapter 5, equation (5.16). In terms of the Method of Stationary Phase, the contribution is still described by the first term in equation (6.7) which will yield nonzero contributions whenever this traveltime difference is smaller than half the wavelet length, i. e., $|\tau_D(\xi^*) - \tau_R(\xi^*)| < \tau_w/2$, as discussed above.

6.2.3 Points on the reflector boundary: M_3

The point M_3 represents the boundary of the migrated reflector image. The Huygens curve of this point is, in principle, equivalent to the one of point M_1. However, as the stationary point is located directly at the margin of the ZO gather, only half the operator is within the data volume. Thus, summing up along the stacking curve results in an amplitude value that is half of the value assigned to M_1. This coincides with the stationary-phase analysis for the case when the stationary point falls on the boundary of the integration interval. As mentioned in connection with equation (6.4), in this situation the leading term is divided by two.

6.2.4 Points off the reflector: M_4

Points such as M_4 represent the majority of diffraction points within the target zone. They have Huygens curves which completely cross the reflection signal. Summing up amplitudes along such operators leads to low values as the result of destructive interference. In other words, the phase of the integrand in equation (6.2) is rapidly varying, so there is no leading order contribution. For the geometrical interpretation of equation (6.7), the point of stationary phase is to be identified with that point in the seismic section where the traveltime and Huygens curves have the same time dip. We see that at this point as well as at its endpoints, the operator lies outside the signal, i. e., $|\tau_{dif}(\xi^*)| > \tau_w/2$, $|\tau_{dif}(a_1)| > \tau_w/2$, and $|\tau_{dif}(a_2)| > \tau_w/2$. Therefore, all terms in equation (6.7) are zero.

6.2.5 Migration artifacts caused by the finite stacking operator: M_5, M_6, and M_7

For points such as M_5, the endpoints of the stacking operator lie within the reflection signal. Because of the limited aperture, the stack does not sum up all the data necessary for complete destructive interference in the same way as it does for point M_4. Thus, the migration output at M_5 is not as low as that for point M_4. As a consequence, a migration artifact appears in parallel to the actual reflector. With increasing size of aperture, the effect at M_5 moves away from the actual reflector and might be located outside of the target zone. Sun (1998) shows that this aperture effect completely separates from the reflector image if the aperture is larger than one Fresnel zone (see also Section 6.4).

The relationship of these observations to the Method of Stationary Phase is straightforward. As for points M_1 to M_4, the point of stationary phase corresponds to the point of equal dip of the traveltime curves τ_D and τ_R. At this point, the stacking line is outside the signal, i. e., $|\tau_{dif}(\xi^*)| > \tau_w/2$. Therefore, the first term of equation (6.7) yields no contribution to the migrated image. However, both endpoints of the operator lie inside the signal, i. e., $|\tau_{dif}(a_1)| < \tau_w/2$ and $|\tau_{dif}(a_2)| < \tau_w/2$. Therefore, the boundary terms of equation (6.7) predict a nonzero migration output at M_5.

The situation at point M_6 is in principle equivalent to that at point M_5. However, because only one endpoint lies within the reflection signal (the other endpoint lies outside the data), the amplitude of the aperture effect at M_6 is just half of that at M_5.

Point M_7 marks the transition between the situations of points M_5 and M_6. The endpoint of its pertinent Huygens curve coincides with the boundary point P in the data, where the survey ends. For this reason, at M_7 the migration artifact splits into two effects. In addition to the limited-operator effect described above, a limited-data effect appears in the migrated traces.

6.2.6 Migration artifacts caused by the finite survey area: M_8 and M_9

The most prominent migration artifact is the migration smile, represented by points M_8 and M_9. The pertinent Huygens curves cross the reflection signal exactly at the end of the survey line. In this way, the destructive interference is incomplete at one of the endpoints, thus leading to a nonnegligible contribution. It is worthwhile to observe that the position of the migration smile is given by the geometrical location of all points of the type of M_8 and M_9 whose Huygens curves cut the border point P of the reflection signal. Note that, because of the duality between the Huygens curve and the isochron (see, e. g., Tygel et al., 1995, and also Chapter 5 and Figure 5.1), this is the isochron of P. The resulting migration artifact follows this isochron, which is a half-circle for my constant-velocity zero-offset experiment as shown in Figure 6.3(a).

Observe the inverted polarity (red is positive, blue is negative) of the artifact between points M_8 and M_9. This can be explained with the help of the symmetry of the operator. The dashed part of the Huygens curve of M_8 that is outside the data is identical to the solid part of the Huygens curve of M_9 that is inside the data. Thus, the stack at M_9 will contribute with exactly that part of the data that is missing at M_8. The actual values of the migration results at points M_8 and M_9 depend on the form of the source wavelet as well as on the filter (i. e., the half-derivative) applied in the migration process. However, the fact that these values are complementary to each other is independent of these conditions. For a better visualization of this complementarity, I pick the peak amplitudes along both branches of the migration smile corresponding to points M_8 and M_9 and add them. We can verify in Figure 6.3(b) that the sum of amplitudes at each pair of two opposite points from the two branches indeed yields zero (except, of course, for a numerical error).

Again, we can directly relate the above geometrical interpretation to the terms of the stationary-phase evaluation of the Kirchhoff migration integral. The migration outputs at points M_8 and M_9 are described by the third term in equation (6.7). The first term yields a zero contribution since, at the stationary point, the stacking line lies outside the reflection signal as in the case of points M_4, M_5, M_6, and M_7. The second term in equation (6.7) which stems from the upper integral limit a_2, is also zero. At both M_8 and M_9, the actual contribution stems from the lower integral limit, $a_1 = 2500$ m. As the Huygens curves of both points terminate at the same position, $f(a_1)$ is the same for both of them. So where is the inverted polarity? It is in the sign of the derivative $q'(a_1)$, i. e., in this simple example the dip of the stacking curve at the survey end. As we can easily observe in Figure 6.2 this sign is positive for M_9 but negative for M_8.

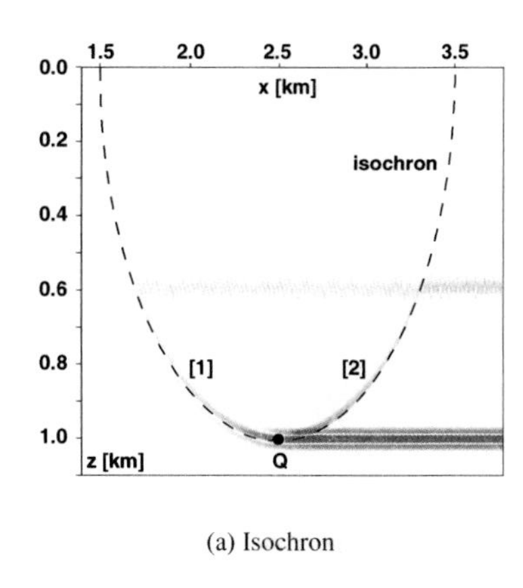

(a) Isochron

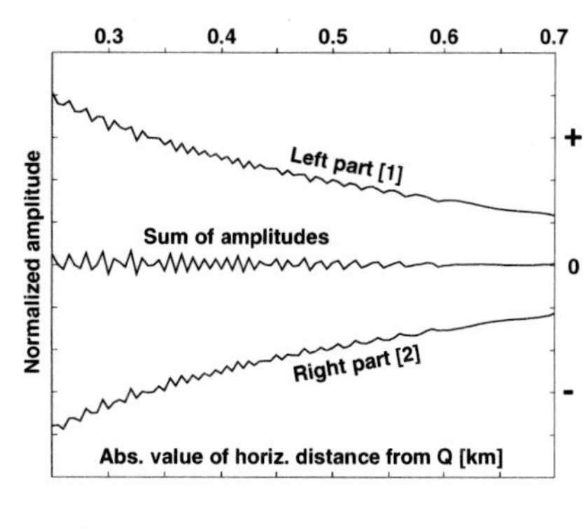

(b) Amplitudes along the artifact branches

Figure 6.3: Analysis of the migration smile. (a) Kinematically, it coincides with the isochron of the border point P of the data. (b) The sum of peak amplitudes of two opposite points on the isochron branches [1] and [2] yields approximately zero.

The Method of Stationary Phase evaluation allows for a more quantitative analysis of the migration smile. Using equation (6.4) and recalling the additional factor $\sqrt{\omega}$ in front of the integral in equation (6.1) (which stems from the time half-derivative in the original Kirchhoff migration integral), we see that the main contribution to the migration result will be frequency independent while the boundary effects will decay proportionally to $1/\sqrt{\omega}$. Figure 6.4 shows the amplitude of the migration output at points M_5 (circles) and M_8 (crosses) as a function of the dominant frequency of the source wavelet used in the modeling. The actually observed amplitudes follow almost exactly the predicted behavior (solid line).

6.2.7 Prestack migration and comparison with Sun (2000)

The reader might notice that the examples of limited-aperture migration (LAM) in Sun (2000) do not distinguish between the different types of migration artifacts described here. The reason is quite simple: Sun uses a prestack migration example with a single shot only and, in addition, he shows only a single trace of the migration result in the center of the survey. In that case, only one artifact is visible, namely that due to the limited operator as represented above by points M_5 and M_6. Of course, both aperture effects are also present in prestack migration, as can be seen in Figure 6.5.

To construct this figure, the data were sorted into common-offset gathers and then migrated separately. The actual migration operator was limited to a maximum aperture radius of 0.8 km around its apex. The respective migration results are displayed in planes parallel to the front face of the cube. In this way, the axis perpendicular to the front face of the cube represents the source-receiver offset. The front face itself is identical to the zero-offset migrated section shown in Figure 6.2. The side face of the cube is an image gather that depicts the same depth-migrated trace for every offset. Since the migration was performed in a true-amplitude sense, the picked amplitude along an event in the image gather would yield the AVO curve for the respective depth point, see also Chapter 7.

As seen from Figure 6.5, both aperture effects vary as a function of offset. The artifact due to the limited operator size shows a moveout in the image gather (the side face of the cube in Figure 6.5). For

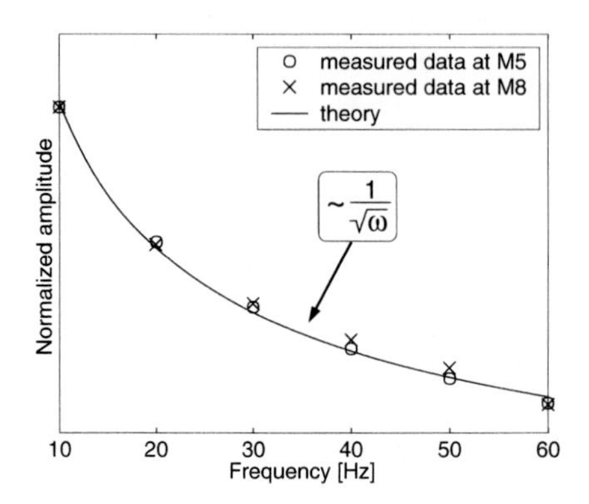

Figure 6.4: Frequency behavior of the boundary effects in 2.5D. The amplitude at M_5 (circles) and M_8 (crosses) decays with $1/\sqrt{\omega}$ as predicted by the Method of Stationary Phase.

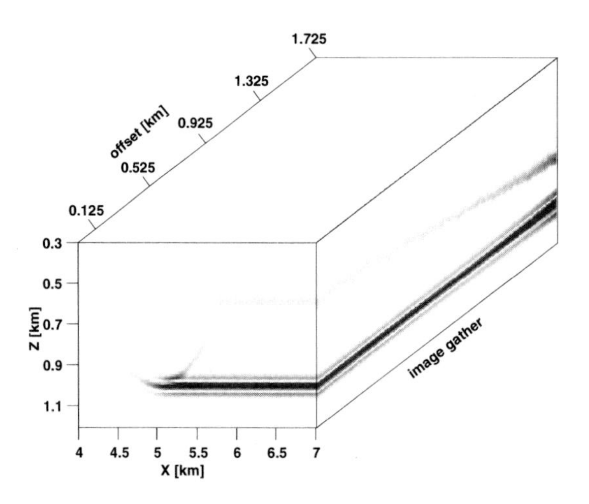

Figure 6.5: 2.5D prestack migration for the same model used in the ZO example. The maximum aperture radius was 0.8 km. Both types of artifacts can be observed. Due to the offset dependence of the operator, the corresponding artifact shows a moveout in the image gather (the side face of the cube). Note the offset-dependent stretch of the wavelet defining the reflector in the depth domain. The stretch occurs due to the increasing reflection angle with offset and is described according to equation (5.16).

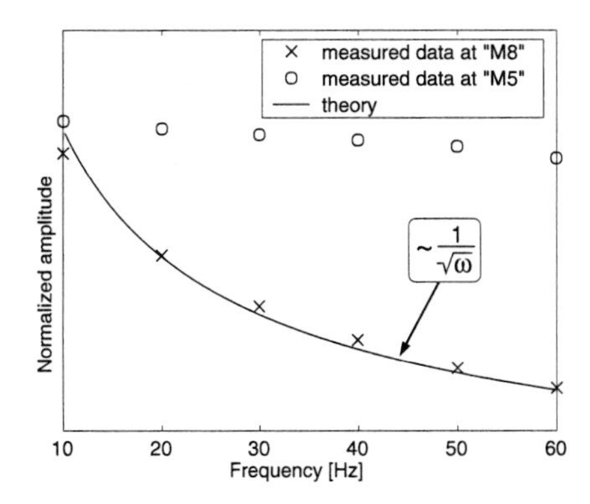

Figure 6.6: Amplitude behavior of the boundary effects in a 3D migration. Circles: amplitudes at a point "M_5", crosses: amplitudes at a point "M_8", solid line: $1/\sqrt{\omega}$ behavior predicted by the Method of Stationary Phase for point "M_8".

larger offsets, this artifact moves closer to the migrated reflection because the curvature of the operator (and, thus, the traveltime difference between the operator endpoints and its apex) reduces with offset. The isochron-type artifact due to the limited survey area (i. e., the migration smile) broadens and moves along the x-axis because the reflector illumination changes with offset. Because of this offset dependence, a postmigration stack can significantly reduce both migration artifacts. In spite of that, in complex media some strong artifacts will generally remain visible in the final migrated section.

6.3 Boundary effects in 3D

In 3D, the physical conditions that cause boundary effects are the same as in 2.5D, these being the limits of the seismic data and of the stacking operator. Therefore, the migration artifacts observed in 3D Kirchhoff migration are conceptually the same as in 2.5D migration. One observes the migration smiles from the survey ends as well as the reflector shadow as a result of the limited operator size. This is confirmed by a corresponding stationary-phase analysis of the Kirchhoff migration integral, which also reveals the two leading-order contributions to be those from the stationary point(s) and the integration limits (see, e. g., Wapenaar, 1992; Sun, 1999).

However, in quantitative terms, the increase in dimension slightly changes the geometrical situation. In 3D, Kirchhoff migration is realized by the double integral (5.2). Consequently, the stacking operator is no longer a line but a surface, and its boundary is not a point but a line. For that reason, the amplitude behavior of the artifacts can be different.

Figure 6.6 shows corresponding numerical results from a 3D migration. The model and all its parameters are the same as for the 2.5D experiment, with identical extension into the third dimension. Indicated is the $1/\sqrt{\omega}$-behavior (solid line) together with the amplitudes of the 3D migration artifacts at points that correspond to points M_5 and M_8 in Figure 6.2, here denoted in quotation marks, i. e., "M_5" and "M_8". The amplitude of the artifact at point "M_8" decays with $1/\sqrt{\omega}$ (as in the 2.5D case). However, the artifact built up by points such as "M_5" shows almost no frequency dependence.

The observed amplitude behavior of both effects can be explained by the 2D stationary-phase evaluation of the 3D Kirchhoff migration integral. The reason for the different trends is that the main contribution at "M_8" comes from a stationary point of the boundary line integral (see, e. g., Bleistein, 1984), while at "M_5", this integral is no longer of oscillatory character and thus contributes almost uniformly over all frequencies. However, it goes beyond the scope of this section to enter into the mathematical details of 3D migration artifacts and to comment on all similarities and differences to the 2.5D situation.

6.4 How to avoid aperture effects

Above, I have already indicated that there is a well-known technique to reduce migration artifacts resulting from the limited migration aperture. All that has to be done is to avoid an abrupt end of the operator but let it die off over a couple of traces, i. e., apply a taper. This has to be done at two different places: firstly, the input seismograms are tapered at the endpoints of the survey area. Secondly, the finite operator is not just truncated but also tapered at its endpoints. In terms of the stationary-phase solution (6.7), the values of $f(a_1)$ and $f(a_2)$ are artificially set to zero. This must be done smoothly to avoid a violation of the underlying assumption of a slowly varying function $f(\xi)$. Then, this approach reduces the contributions of the operator endpoints and, thus, helps to obtain a migrated image with less migration artifacts.

6.4.1 Size of the taper region

When applying a taper, the fundamental question is over how many traces it should extend. On the one hand, the taper ought to be large enough not to violate the smoothness assumption so as to effectively suppress the artifacts. On the other hand, it should not be too large so as not to lose more information than necessary on the amplitudes at the survey ends or to stack unnecessary information at the operator ends. Sun (1998) suggests that in the same way as the stacking region should cover the first (projected) Fresnel zone, the taper region should extend over the second (projected) Fresnel zone around the stationary point. Unfortunately, this point cannot be estimated prior to or during migration. Therefore, we must once again compromise to avoid the aperture effects.

To get an idea about the size of the taper region, I propose the following simple criterion for zero-offset (poststack) migration. As is well-known, to kinematically migrate all reflectors at depth z up to maximum dip angle θ_m, the stacking operator may be restricted to a radius of

$$r = z \tan \theta_m \; . \tag{6.9}$$

If the same reflectors are to be migrated dynamically correctly, the radius must be increased by the size $FZ(1)$ of the projected first Fresnel zone. As shown in Appendix B, $FZ(1)$ is given in the frequency domain by

$$FZ(n) = \frac{\sqrt{\frac{vznT}{2\cos\theta_m} + \left(\frac{nvT}{4}\right)^2}}{\cos\theta_m} \tag{6.10}$$

with $n = 1$, where v is the medium velocity and T the period of the considered monofrequency wave. As in equation (6.3), the half-period $T/2$ has to be replaced by some estimate of the wavelet length τ_w if formula (6.10) is to be applied in the time domain. According to Sun (1998), the artifacts are suppressed as well as possible, while affecting the amplitudes as little as possible, when the operator is increased by $FZ(2)$ instead of $FZ(1)$. The additional operator extension $FZ(2)-FZ(1)$ is the second

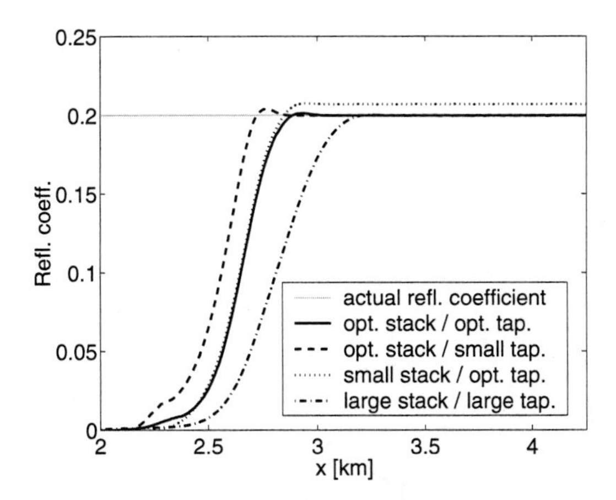

Figure 6.7: Comparison of migrated (zero-offset) amplitudes for different sizes of the aperture and taper region. The synthetic ZO section shown in Figure 6.2 was used. The optimal taper size is the one that eliminates all artifacts but recovers the amplitudes as close to the margins as possible.

projected Fresnel zone over which the taper is to be applied. Of course, the formulas (6.9) and (6.10) above are strictly valid for constant velocity, only. For inhomogeneous media, they can only be used as a rule of thumb to get a rough idea about the aperture size and the taper region.

Formula (6.10) can also be used to obtain an estimate for the size of the end-of-survey taper. By substituting $z = vt \cos\theta_m/2$ and setting $n = 1$, the size of the taper at two-way time t can be estimated. If a constant taper size is preferred, t can be replaced by the maximum time value in the data. Correspondingly, z in equation (6.10) can also be replaced by the maximum depth in the desired migrated image.

Figure 6.7 compares the amplitudes along the reflector image for different combinations of aperture and taper sizes for the 0° migrated data from the synthetic ZO section of Figure 6.2. When the aperture is too small, not even the amplitudes far away from the data margins are correctly recovered (dotted line), although the optimal taper is used. When the optimal (or a larger) aperture is applied, all amplitude problems are restricted to the data margins. For too small a taper, the survey-end artifact is not completely removed (dashed line). Too large a taper destroys the amplitudes where they can be retrieved from the data (dash-dotted line). The optimal taper size is the one that eliminates all artifacts but recovers the amplitudes as close to the margins as possible (solid line).

The taper function used for the migration examples shown here is a two-sided Hanning window for both the operator and the end-of-survey taper. For comparison, I also tested a two-sided triangular window. The shapes of these functions are depicted in Sun (1998, 2000) for 2D and 3D. Both types of taper functions yield nearly identical results. The optimal values for the aperture and taper sizes were calculated by means of equation (6.10) with $z = 1$ km, $v = 2$ km/s, $\tau_w = 50$ ms, and $\theta_m = 0°$, resulting in $FZ(1) = 320$ m and $FZ(2) = 458$ m. To test the effects of too small or too large aperture and/or taper sizes, the stacking region $FZ(1)$ and the taper region $FZ(2) - FZ(1)$ were halved or doubled, respectively.

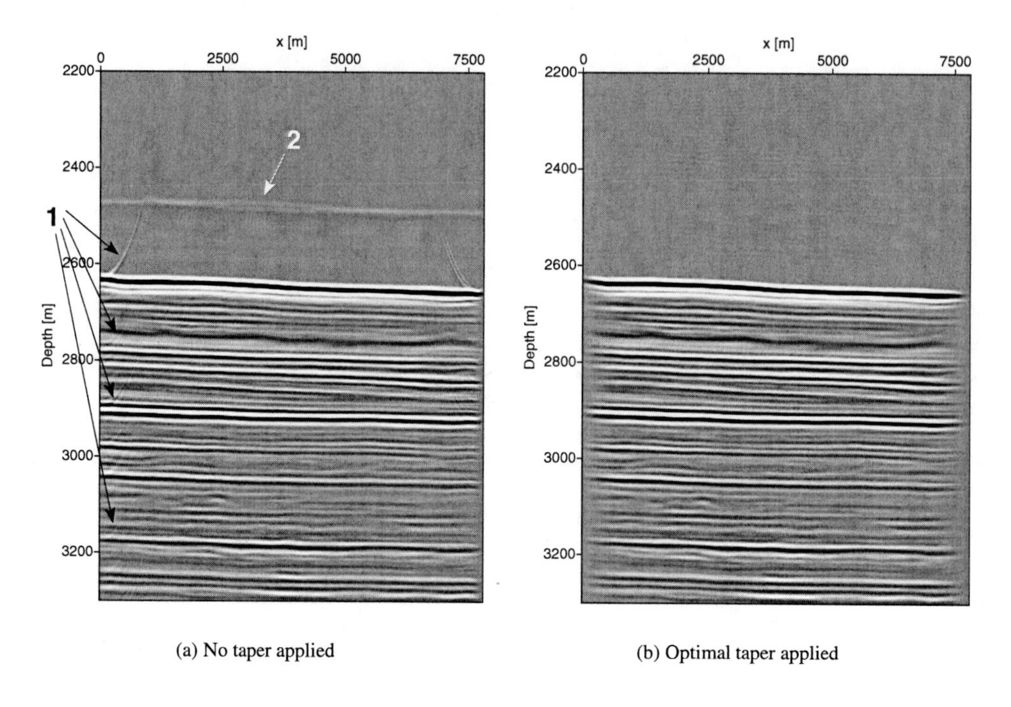

(a) No taper applied

(b) Optimal taper applied

Figure 6.8: Effects of tapering for a real data example. Please note the scale of the vertical axis. (a) Migration result without tapering. The artifacts are denoted by the numbered arrows. 1: artifacts due to the limitation of the survey. 2: artifacts due to the truncation of the stacking operator. (b) Migration result with optimal taper function applied according to equation (6.10). Artifacts are suppressed without loss of information.

6.5 Data example

Figure 6.8 demonstrates the effect of tapering applied to the input data and the stacking operator for a marine data set from offshore British Columbia, Canada (Scheidhauer et al., 1999; Rohr et al., 2000). The image was obtained by a poststack depth migration of 626 traces with a CMP spacing of $\Delta\xi = 12.5$ m and time sampling interval of $\Delta t = 4$ ms.

Figure 6.8(a) shows the migrated reflector image when stacked with a dip-limited 10° migration operator using the optimal aperture of one estimated projected Fresnel zone according to equation (6.10) without applying a taper. Both the migration artifacts due to the limited operator and survey area are present, as indicated by arrows. We immediately recognize the artifacts due to the limited survey aperture that occur at the left and right margins of the picture and are marked by (1). The artifact marked by (2) is the reflector shadow due to the limited migration operator. It is parallel to the actual reflector images and may lead to a misinterpretation of the final migration result. It should be kept in mind that, although only the reflector shadow of the sea bottom is clearly visible in Figure 6.8(a), similar artifacts parallel to all other reflector images are also present, even if hidden by other reflectors. Since the migration was performed in a true-amplitude sense, the amplitudes of the migrated reflection events are expected to be proportional to the zero-offset reflection coefficient. The hidden reflector shadows, however, may compromise the amplitudes and hamper further amplitude analyses of the migration result.

Figure 6.8(b) shows the same migrated reflector image with the optimal taper according to equation (6.10) applied. Both artifacts are almost completely eliminated without affecting the amplitude of the reflector image. Note that also the hidden reflector shadows are eliminated.

6.6 Some general remarks on tapering

At this stage, let me point out that with respect to tapering, I algorithmically agree but conceptually disagree with Sun (1998, 2000). As opposed to him, I do not think the taper function should be conceived of as a part of the weight function because of the following reasons: firstly, in kinematic Kirchhoff migration schemes there exist no true-amplitude weight functions. However, taper functions are still required to obtain a high-quality migration result with reduced artifacts. Secondly, there are *two* taper functions that need to be applied. One serves to avoid the aperture effect of the limited survey area. This taper is completely independent of any weight function and applied directly to the input data before migration. The second taper is applied to the operator during migration and may be implemented as a part of the weight function. I stress once more that the mentioned disagreement is rather conceptual than technical. I prefer to think of the true-amplitude weight and the taper functions as different concepts, even though they may be combined in practice to speed up the algorithm.

6.7 Summary

In this chapter, I have provided an intuitive explanation of aperture effects by discussing the constructive and destructive interference of the stack in simple geometrical situations. This helps to relate the terms of the stationary-phase approximation to the actually observed migration artifacts. It turned out that for practical applications one has to distinguish between two principal types of artifacts. These are

- boundary effects due to a limited survey aperture, and
- artifacts due to a limited migration operator.

Both types of artifacts are mathematically equivalent and can be explained by means of the boundary terms that result from the stationary-phase analysis of the migration integral. Based on the geometrical analysis, I had a closer look at a well-known way to avoid the aperture effects: tapering. The most important question with respect to tapering is how to determine the taper region. Too small a region will not suppress the effects while too large a region will destroy more information than necessary. I have shown that the ideal taper region is closely connected to the minimum aperture. Schleicher et al. (1997) have derived the minimum aperture for a dynamically correct migration to be the first projected Fresnel zone around the stationary point. Sun (1998) has demonstrated that the same minimum aperture of the size of the first projected Fresnel zone is sufficient to separate the operator-end effect from the desired image. I have confirmed both observations numerically. Moreover, to avoid the operator-end effect, a taper region of the size of the second projected Fresnel zone should be added to the operator. In principle, the projected Fresnel zone(s) can be determined during migration, even in inhomogeneous media, from dynamic ray quantities. However, to speed up the process, it is often useful to fix the operator size beforehand. As I have demonstrated with a real-data example, the constant-velocity formula helps to get an idea of an adequate aperture and taper region.

In the next chapter, further aspects of (Kirchhoff) migration are discussed.

Chapter 7

Further aspects of (Kirchhoff) migration

As we have seen, migration is an important step in the processing of seismic reflection data. It is necessary to obtain a clear image of the subsurface where reflectors are correctly positioned. In addition, migration might serve as an intermediate step to create input data for further lithostratigraphic analyses. In the introduction it was pointed out that there exist numerous methods and a lot of different implementations to achieve this goal. Due to its underlying geometrically appealing description, Kirchhoff migration is one of the oldest but still widely investigated methods in seismic migration. Although some different migration approaches have been developed in the last decades that might produce better images of the subsurface (at least in some scenarios), Kirchhoff migration is still a competitive and frequently used tool. The question is why? To answer this question, let me briefly summarize some features of Kirchhoff migration:

- The kinematic (i. e., traveltime-related) aspects of the method are easy to understand and describe. This was explained in detail in Chapter 5.
- There exists a firm mathematical treatment which provides the basis for Kirchhoff migration, see Chapter 4. This mathematical foundation was developed by Schneider (1978) and other scientists in the following years. The theory was later extended to provide not only kinematically but also dynamically correct images of the subsurface. Bleistein (1987) was one of the first dealing with the theoretical formalism of what is today know as true-amplitude migration, see also Chapter 5. Hubral et al. (1996) and Tygel et al. (1996) provided a complete unified theory to perform 3D true-amplitude seismic reflection imaging based on Kirchhoff migration and its asymptotic inverse process called Kirchhoff demigration.
- Kirchhoff migration is able to handle vertically and laterally inhomogeneous media and provides (in most cases) reliable and accurate results while being efficient at the same time. It remains less computer intensive than other methods.
- The method itself is very flexible and allows target-oriented processing of seismic data, i. e., only a selected domain in space can be imaged. This is very often advantageous in a model-building process: imaging a single line, imaging a subset volume below a given depth, or computing a single image gather is not only possible, but even common practice.
- Measurement surfaces with varying topography and, closely related, irregular acquisition geometries can be readily handled without the need for redatuming or data regularization. Fur-

thermore, the method can handle turning rays—however, this is also a question of the method utilized to calculate the relevant Green's functions.

Kirchhoff migration has, of course, also some drawbacks that are outlined below. Special aspects will also be discussed in more detail in later sections.

- Most implementations of Kirchhoff migration use an asymptotic high-frequency and a far-field approximation (Chapter 4). Such approximations cause problems for the region that is within several wavelengths of the source or receiver position, i. e., we may face problems when imaging near-surface reflectors. However, this is only a minor problem in practice and even for very shallow reflectors, Kirchhoff migration is able to produce reasonably good images. Nevertheless, all limitations of ray tracing in complex media remain, especially the very high sensitivity of traveltimes to the model. Problems are usually avoided by using fairly smooth macrovelocity models, although it may be at the cost of the image quality.

- Multipathing may occur depending on the complexity of the macrovelocity model. Most Kirchhoff migration implementations assume that there is only one possible path from a diffraction point M in the subsurface to a source or receiver point at the measurement surface. There exist some extensions that can handle the multipathing problem in Kirchhoff migration, at least for a limited amount of raypaths (usually up to three). However, this comes along with a noticeable increase in computation time. The multipathing problem may also be addressed by means of Gaussian beam migration (Červený et al., 1982; Hill, 1990, 2001; Popov, 2002) that uses a different approach than standard Kirchhoff migration but with similar flexibility. However, it is also associated with increased computational effort.

- Kirchhoff migration sometimes fails to image complex structures (e. g., reflectors beneath a salt body). This is closely related to the traveltime tables used in the migration process. The accuracy and the quality of the migrated image strongly depends on the method used to calculate the Green's function. There exist a lot of methods and even more implementations to generate the traveltime tables (e. g., simple eikonal solvers or kinematic and dynamic ray tracers) which expose different speeds and accuracies, see also below.

- The operator aliasing problem is significant in Kirchhoff migration as we sum up discrete data along a diffraction surface without regarding the frequency content. It may happen that the steeper parts of the Huygens surface undersample the wavelet and, thus, the migrated image severely suffers from aliasing effects. Abma et al. (1999) and Biondi (2001) give a profound description of the problem and its possible solutions. The topic is also discussed in a later section.

- The computational costs for the estimation of true-amplitude weights in Kirchhoff migration are, in general, high because several dynamic ray quantities are involved that must be computed in addition to the traveltime itself, see Chapter 5. However, there exist some approaches to overcome this problem. They are either based on simple approximations of the complete true-amplitude weight factor (e. g., Peres et al., 2001) or on the determination of the weight function from traveltimes only (Vanelle and Gajewski, 2002b,a; Vanelle, 2002).

As shown by Gray (1998), most migration methods are accurate when imaging regions of typically observed structural complexity (i. e., if one ignores very complex salt structures, etc.). The differences are usually less than the uncertainty in estimating the velocity model. Therefore, using much more expensive tools to only slightly increase the accuracy is mostly not justified from an economical point of view. Because of its accuracy and relative cheapness, Kirchhoff migration is a workhorse in seismic data processing.

7.1 Green's function calculation

As we have seen in Chapter 4 from a mathematical point of view and in Chapter 5 from a geometrical point of view, the Green's functions used in seismic imaging are functions representing the propagation of waves from one point in the subsurface to another one through a given or estimated velocity model. One of those subsurface points is the point to be imaged, the other one a point at the surface, i. e., a source or receiver position. In this way, the Green's function acts as a transfer function between these points in our imaging method. The Green's function can be decomposed into two terms, one related to traveltimes, the other related to amplitudes. Therefore, also the calculation methods used to estimate the Green's function can be divided into two groups: one with methods that calculate traveltimes, only, and the other with methods that calculate the full (ray-theoretical) Green's function. An application of a method of the former type assumes in general that amplitudes are handled separately, usually in an empirical or semi-empirical manner during preprocessing, whilst the second group is considered more precise because amplitudes are dealt with deterministically. However, the additional calculation effort required needs to remain compatible with the demands of available computing power and the time required for processing.

The majority of calculation methods use ray-tracing techniques. These are purely kinematic in the first class of methods, and dynamic for the second class. Alternatively, an integration of the eikonal equation (3.4) might be used (eikonal solver, e. g., Vidale, 1988). Two-point ray tracing, although very comprehensible, should be avoided because this requires an undesirably large number of iterations and is, thus, inefficient. In contrast, an efficient method to create Green's function tables (GFT) for migration is, e. g., the *Wavefront Construction Method* (Vinje et al., 1993, 1996a,b) where complete wavefronts are propagated through a medium either from a point at the surface or from a point at depth. Unfortunately, even with those sophisticated tools at hand, the number of calculations and the required storage capacities are enormous. Hence, certain simplifications are made. One of the most commonly used is to calculate the Green's functions using a coarse grid (with the result being stored as Green's function table), even though these files are still of considerable size. The traveltime parameters and weights required in the migration process are successively derived, as needed, from the tabulated values by means of interpolation, see, e. g., Vanelle and Gajewski (2002a).

The precision with which the propagation velocity field is estimated and the Green's functions are calculated will directly affect the quality of the migration result. Generally, the greater the precision in the velocity field, the more complex it is and, consequently, the more difficult and costly the calculation of the Green's functions. Therefore, the estimation of a suitable macrovelocity model for depth imaging from the measured data and the rapid and precise calculation of the Green's functions in any medium (including, e. g., also anisotropic media) form the keys in industrial 3D seismic imaging.

The multipathing problem was already mentioned above. In current practice, often only one solution for the Green's function (and, thus, only one possible raypath between two points in the subsurface) is used. The representation is usually the most energetic (the preferred) or the shortest time (easiest to compute) solution. However, especially in complex areas, multipathing must be considered in

Kirchhoff migration in order to obtain a high-quality image of the subsurface. If other methods are used for migration, e. g., finite-difference wave equation migration, then multipathing is implicitly taken into account, but with the drawback of dramatically increased computation time.

A summary of available methods and tools to calculate the Green's function and a data example based on the complex Marmousi model is presented by Audebert et al. (1997).

7.2 AVO and AVA issues

In the introduction and also in later chapters, it was mentioned that in prestack (common-offset) depth migration all offset panels are migrated separately, thus yielding a complete subsurface image for each offset (see Figure 1.7). Common-image gathers (CIG) are formed by collecting all those traces that represent the same lateral position but belong to different migrated common-offset panels. As is explained in detail in the next section, image gathers allow an assessment of the validity of the velocity model and, hence, the possibility of velocity analysis (e. g., Jin and Madariaga, 1993, 1994). Moreover, there is an additional benefit if the offset dimension is retained in prestack depth migration. If a true-amplitude migration has been performed, this allows to study the reflectivity as a function of offset, which is precisely the goal of an amplitude-variation-with-offset analysis (e. g., Beydoun et al., 1993). I assume here that the image gathers are flat, i. e., the velocity model underlying the imaging process is correct.

For a wave incident upon an interface, the amplitude of the reflected wave will vary as a function of the angle of incidence as well as the medium parameters above and below the interface. This is the phenomenon that forms the basis of the amplitude-variation-with-offset (AVO) or amplitude-variation-with-angle (AVA) analysis, respectively. Such analyses allow to gain a better understanding of rock properties on either side of the interface compared to the reflectivity data at normal incidence alone. A law governing the variation of reflectivity with angle of incidence for a planar interface separating two isotropic homogeneous media was established by Zoeppritz (1919). Several simpler approximations have since been proposed, one widely used being that of Shuey (1985):

$$R_{PP}(\alpha_{M_R}) = R_{PP}(0) + G \sin^2 \alpha_{M_R} \, , \tag{7.1}$$

where R_{PP} is the reflection coefficient for a monotypical P-wave reflection, α_{M_R} is the reflection angle, $R_{PP}(0)$ is the normal incidence reflection coefficient, and G is a linear parameter depending on $R_{PP}(0)$, P-wave velocities, densities, and the Poisson ratio. The parameter $R_{PP}(0)$ is also known as the *intercept*, while G is denoted *gradient*; both are estimated by linear regression with respect to the parameter $\sin^2 \alpha_{M_R}$, see Figure 7.1.

If true-amplitude weights are correctly estimated within the Kirchhoff migration process, we are able to extract intercept and gradient from the image gathers after prestack depth migration. The offset is directly (but not necessarily uniquely) linked to the angle of incidence of the reflection upon the reflector. This means, for each imaged position M_R, we can use, for instance, a ray-tracing tool, once the image has been interpreted, to determine the angle of incidence for given offsets at each point involved. Repeating this operation for every offset, a relationship can thus be established between the measured reflectivity and the specular reflection angle. This is the aim of AVA analyses. During the last years, algorithms have been developed to directly obtain common-angle images after prestack depth migration, see, e. g., Xu et al. (2001). This is extremely useful for direct AVA analyses, but requires more computational expenses.

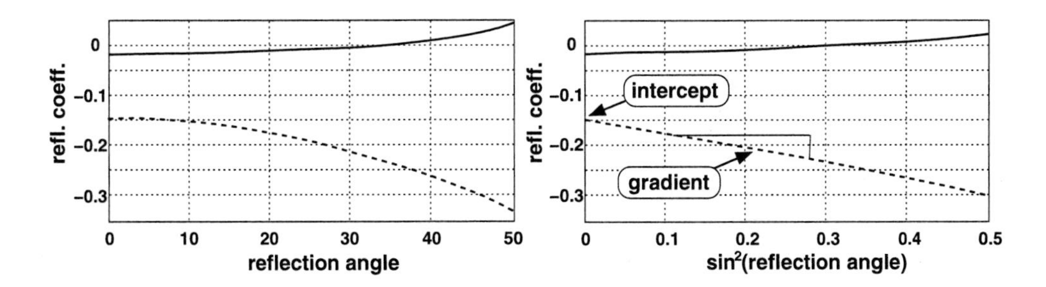

Figure 7.1: Amplitude-variation-with-angle (AVA), here indicated for a water-saturated sand over an oil- (solid line) or gas-saturated (dashed line) sand. The left picture shows R_{PP} vs. α_{M_R}, the right picture shows R_{PP} vs. $\sin^2 \alpha_{M_R}$. In this way, the intercept and gradient can be estimated by linear regression according to Shuey (1985). Note that the curves in the right picture are only strictly linear up to moderate reflection angles.

Whatever domain is considered, either AVO or AVA, we need precise amplitudes for an analysis to be meaningful. This is one major task of true-amplitude imaging. In addition to high-quality seismics, we usually need well data to calibrate the results. An AVO or AVA analysis is, at best, difficult and, at worst, impossible; but when it works, it can reveal much more information than that (apparently) present in the migrated section.

7.3 Velocities

It was mentioned several times that an appropriate macrovelocity model is needed in order to migrate the data and, thus, transform it from the time to the depth domain. While for time migration a root-mean-square (RMS) velocity model is sufficient, depth migration explicitely requires local (interval) velocities. The quality of the model that needs to be estimated prior to the migration process strongly influences the quality and reliability of the migrated image. Velocity analysis and the velocity model building process are, therefore, an important part in seismic reflection imaging. Three main classes of models are commonly used in (prestack) depth migration projects:

- Layer-based models, where layers with very different velocity regimes and thicknesses are overlaid one on top of the other,
- gridded models, where the velocity is usually defined at each node of a regularly spaced grid (this may, of course, be implemented in a different way), and
- hybrid models where a fairly smooth velocity-gradient model is complicated by the presence of massive bodies (e. g., salt or shale) of a complex shape with ragged surfaces and high velocity contrasts.

In practice, the velocity model used in prestack depth migration can be only relatively simple, but should nevertheless reflect as accurately as possible the actual traveltimes in the subsurface in all directions in space. This condition is all the more important, although a great deal more difficult to achieve, when the medium is strongly heterogeneous and/or anisotropic. Which type of model to use is sometimes difficult to decide and the velocity model building process itself is often an integral part of (prestack) depth migration.

A vast majority of velocity analysis methods is iterative, where iterations take place layer by layer or as updates of an initial model. Several different methods are available. Among these are, for instance, focusing analyses, common focus point (CFP) analyses, 1D analytical updates ("Deregowski loop"), traveltime inversion, stereotomography, picking and inversion of common-image gathers, velocity scanning, or statistical searches (Robein, 2003). In all update methods, the initial model is very important; the closer it is to the true effective medium, the more likely the migration will be completed in a reasonable amount of time. Update techniques actually assume small corrections and might fail if the initial model is too far from the true effective model. Usually, one uses an economical method (see, e. g., Yilmaz, 2001; Robein, 2003) to estimate the parts of the model with limited velocity variations, and subsequently applies (iterations of) prestack depth migration to refine the model by concentrating on improving the image within and below the zone of complexity. The most classical criterion to assess whether a model is the optimal one for prestack depth migration and consistent with the data is the following: a necessary condition is the flatness of events in a common-image gather. If an analysis of events in an image gather yields that they are not flat, we must correct (i. e., update) the velocity model and perform the prestack migration process again. The idea of inverting image gathers was first published by Al-Yahya (1989) who proposed an approach within the prestack shotpoint migration domain. The principles remain valid for image gathers calculated using common-offset migration with some modifications to the formulas. Based on a homogeneous model description of the subsurface, Al-Yahya (1989) derived an analytical formulation of the theoretical shape of events in the image gathers as a function of the velocity in the initial model. He showed that this shape

- curves upwards if the overall velocity was too small, i. e., images from large offsets appear at shallower depths than those from shorter offsets;
- bends downwards if the overall velocity was too large.

The shape of events in the image gathers is defined by a parameter γ that is estimated by means of a semblance-based analysis. The difficult task then is to transform an estimate of γ into an update of the velocity model. This is usually achieved by means of a least-mean-squares optimization process. Al-Yahya's method has been further developed, see, e. g., Billette et al. (2000). The method provides an efficient way of building a model with limited complexity but will fail in complex tectonic situations due to its inherent assumptions. One way to overcome this problem is the use of ray-based image gather inversion. Such schemes have been applied industrially in 3D for either isotropic or anisotropic models (Williamson et al., 1999).

7.4 Aliasing in Kirchhoff migration

Aliasing is a common problem in seismic data processing because data is recorded in a discrete way both in time and space. The related aliasing problem is referred to as *data aliasing* and is independent of the method utilized to process the data. Data aliasing can only be avoided by properly planning and executing the data acquisition, i. e., a proper temporal and spatial sampling is needed to handle the frequency content of the data and the dips to be expected, see, e. g., Yilmaz (2001). It is well known that the maximum frequency contained in the data must be smaller than the Nyquist frequency,

$$f_{Nyq} = \frac{1}{2\,\Delta t}\,, \tag{7.2}$$

in order to avoid temporal aliasing, where Δt is the time sampling of a seismic trace. A corresponding equation holds for the Nyquist wavenumber and spatial aliasing, see also Vermeer (1990).

In the output domain, we may face *image aliasing* which occurs if the spatial sampling of the image is too coarse to adequately represent the steeply dipping reflectors that the imaging operator attempts to build up in the imaging process. Such kind of aliasing can easily be avoided in Kirchhoff migration by using appropriate (i. e., sufficiently dense) output grids. Biondi (2001, 2003) discusses image aliasing and shows how to avoid image aliasing if the output spatial sampling is fixed and cannot be adjusted.

In addition, Kirchhoff migration faces a special kind of aliasing related to the operator that links the data space with the image space. If this operator (i. e., the Huygens surface) is not carefully implemented, artifacts are generated when data that are not aligned with the summation path stack coherently into the image. In other words, it may happen that the steeper parts of the Huygens surface undersample the wavelet of adjacent traces and, thus, cause artifacts known as *migration noise* or *aliasing noise* to appear in the output image. This phenomenon is called *operator aliasing*. Note that operator aliasing may occur even when the data are not aliased at all and can severely degrade the migrated image. Therefore, it is necessary to account for operator aliasing in Kirchhoff migration. Anti-aliasing requires band-pass filtering the data as a function of the operator dip. If it is not efficiently implemented, this band-pass filtering can be expensive. Thus, different approaches exist to handle the problem.

As was explained in detail in Chapter 6, a restriction of the stacking operator to a certain spatial extension or a certain dip can help to avoid the operator aliasing. However, this comes along with a maximum reflector dip in the subsurface that can be imaged. If we have no information about possible reflector dips in the subsurface prior to the migration process or if large dips are to be expected (e. g., the flank of a salt dome), we cannot simply restrict the spatial extent of the stacking operator as this would prevent the imaging of important stratigraphic features in the migration target zone. One way to overcome the problem of operator aliasing is the interpolation of (artificial) traces between measured ones because this reduces the distance of adjacent traces and, thus, circumvents that the wavelet is undersampled during stacking (Spitz, 1991). However, the correct interpolation of seismic traces is a difficult topic and requires, in general, a sufficiently high signal-to-noise ratio and identifiable events. Moreover, interpolation of traces in a multicoverage dataset is a very time-consuming process. Thus, one commonly uses anti-aliasing filters (AAF) in order to get rid of the operator aliasing problem.

In principle, one has to filter the input data with a high-cut filter to remove the high frequencies at steep dips of the Huygens surface. Each input sample will contribute to many different positions on many different Huygens surfaces, each of which requires a different high-cut filter. In most implementations, a convolution form of the high-cut filter is used where the length of the filter is proportional to the inverse of the high-cut frequency. There exist many filter options and filter types that may range from a simple box car shape or triangular shape to a more complicated sinc-type shape. The choice of the filter depends on the acceptable processing time and the quality that is required. In addition, the spectral shape of these filters must be considered with respect to the way they truncate the end of the pass band. The desired sharp cut-off frequency of a sinc-type filter will preserve all the energy in the pass band and attenuate all the aliased energy. However, it may require an order of magnitude more processing time than a simple box car filter. As a consequence, the filter design is a tradeoff between computation time and the quality of the filter process, i. e., the amount of aliased energy that is removed compared to the amount of desired energy that is preserved. An efficient triangular filter that can be applied during the migration process on-the-fly was proposed by Claerbout (1992a,b) and later by Lumley et al. (1994). Gray (1992) presented a similar approach but a different implementation. He proposed creating multiple versions of each input trace filtered with various high-cut frequencies. The parts of the migration operator with slopes that produce aliasing above a particular frequency use a version of the input trace with the aliased frequencies removed. If both methods are compared, one recognizes that Gray's approach has a higher setup cost for each input trace than does triangle smoothing on-the-fly. Hence, if each input trace contributes to a large number of output points, Gray's method will be faster than triangle smoothing; but for migration with few output traces, triangle smoothing may be faster. A combination of triangle smoothing and Gray's method may benefit from the strengths of both techniques and should probably be preferred.

Most implementations of frequency-dependent anti-aliasing assume that events of interest are nearly horizontal, i. e., they are primarily designed for Kirchhoff time migration. The high-cut frequency f_{cut} for 2D data is then calculated according to

$$f_{cut} = \frac{1}{2\frac{\partial t}{\partial \xi}\Delta\xi}, \tag{7.3}$$

where $\Delta\xi$ is the input trace spacing and $\partial t/\partial\xi$ is the local migration operator slope (Lumley et al., 1994). For 3D data, equation (7.3) is applied in both the ξ_1 and ξ_2 direction independently. Abma et al. (1999) propose a slightly different criterion to estimate the high-cut frequency, because the approach of Lumley et al. (1994) is shown to be inappropriate for slow velocities and nonzero offsets as larger offsets are overfiltered and, thus, significant energy is lost, resulting in unreliable AVO analyses. Their criterion reads for 3D data

$$f_{cut} = \frac{1}{2\frac{\partial t}{\partial \rho}\Delta\rho}, \tag{7.4}$$

where $\Delta\rho = \max(\Delta\xi_1 \cos\theta, \Delta\xi_2 \sin\theta)$ with θ being the azimuth of the operator gradient with respect to the inline direction. The parameter $\Delta\xi_1$ is the inline trace spacing while $\Delta\xi_2$ is the crossline trace spacing. This updated limit for anti-aliasing high-cut frequencies proved to be reasonable. Zhang et al. (2001) derived the criterion directly from the integral formulation of Kirchhoff migration and provide a unified framework for discussing also several other issues of anti-aliasing, e. g., anti-aliasing in the image space or in beam migration. Note that the migration operator slope is used in equation (7.3), not the slope of the response of a reflection event to the operator. Ideally, one would like to use the slope of the shifted reflection event for $\partial t/\partial\xi$, which would allow the correction for nonhorizontal events, as pointed out by Crawley (1996). However, such an approach requires the reflector dips to be known prior to the migration process. Baina et al. (2003) propose a new criterion explicitely designed for Kirchhoff depth migration, i. e., for general inhomogeneous velocity models. On the one hand, this approach requires a local slant stack (Chauris et al., 2002) to find the optimal anti-aliasing protection and needs, thus, additional computational effort. On the other hand, examples show that this approach has significant advantages, especially with respect to imaging of steeply dipping reflectors. Biondi (2001, 2003) describes how anti-aliasing can be further tweaked in order to perform "Kirchhoff imaging beyond aliasing".

It should be stressed that anti-aliasing filters and true-amplitude migration are somehow contradictory. Zhang et al. (2001) show the amplitude attenuation caused by anti-aliasing and state that AAF's affect the analysis of amplitudes in migrated gathers often more than the choice of the migration weight. Consequently, one has to take extreme care about amplitudes when performing further analyses on the output of the migration process. However, anti-aliasing is often needed in order to obtain a clear image of the subsurface at all (especially of steeply dipping reflectors). Thus, all Kirchhoff migration programs usually account for anti-aliasing. Further aspects and a general overview of factors affecting the frequency content in imaging processes is presented by Jones and Fruehn (2003).

7.5 Topography and irregular acquisition geometries

Exploration and development activities are increasingly applied to more difficult areas showing significant surface topography. Seismic data recorded in such areas impose severe problems to standard processing schemes. In order to obtain high-quality migrated images, an appropriate handling of topographic effects is substantial. Standard algorithms often fail to do so as they generally anticipate data as if acquired on a flat measurement surface with regular acquisition geometries. Static corrections are routinely applied before migration in order to correct for elevation differences between individual shots and receivers. Unfortunately, this method is very inaccurate in case of non-vertical raypaths in the near-surface area and additionally requires a well-known near-surface velocity. Improperly applied static corrections might severely affect the accuracy of the migration result. Although there exist some more advanced redatuming methods that are based on the wave equation (see, e. g., Bevc, 1997), it might be advantageous or even mandatory to migrate data directly from topography. Gray and Marfurt (1995) pointed out that using migration weights that refer to the recording surface yields significant better migration results than using weights referred to a flat datum. Moreover, they showed that datuming does not allow the same level of precision than processing directly from topography. This is all the more important for true-amplitude imaging.

In order to migrate data recorded on a non-flat surface, several aspects have to be considered: migration weights must refer to the actual topographic measurement surface and its local dip and need to consider the local acquisition geometry. In this context, irregular distribution of sources and receivers along the aperture which is very likely when dealing with complex topography has to be strictly honored in the migration algorithm. Kirchhoff true-amplitude migration provides a suitable tool to account for both, distortions caused by a varying surface topography and irregular acquisition geometries. General aspects of true-amplitude migration from topography were presented by Jäger et al. (2003). A detailed derivation of true-amplitude weights that refer to a measurement surface with topographic variations, an implementation, and the application to several synthetic data examples were presented by Spinner (2003).

A question that arises when recording data on a non-flat surface is, whether it is favorable to fix the distance between receiver groups to a (nearly) constant value Δg measured along the topography or if it is better to fix their horizontal increment $\Delta\xi$. If we fix the distance along the topography by Δg, the horizontal increment $\Delta\xi$ will be irregular but always less or equal to Δg. For this reason, the depth-migrated images will in general have less aliasing noise (Gray et al., 1999). Of course, we always need some kind of binning.

In all practical implementations of Kirchhoff migration, the integrals in equation (5.2) for 3D or the integral in equation (5.20) for 2.5D are realized by summations. In 2.5D, the variable of integration $d\xi$ becomes the discrete $\Delta\xi$ which enters into the amplitude that is assigned to the depth point M as

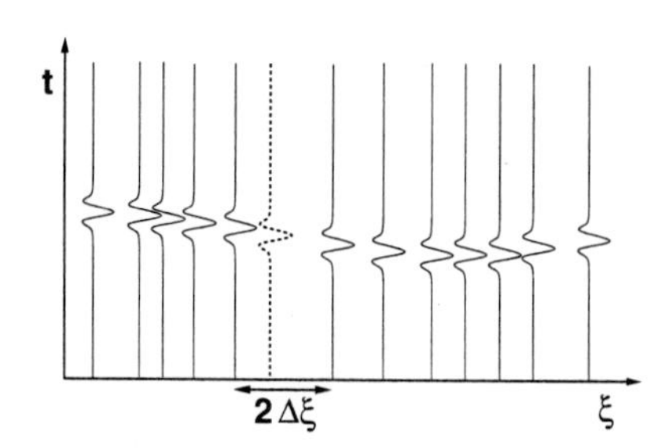

Figure 7.2: Calculation of a local trace weighting factor (for the trace shown as dashed line) corresponding to $\Delta\xi$ in 2.5D poststack migration by means of two neighboring traces.

migration result,

$$V(M) = \sum \Delta\xi \, W_{DS}^{2.5D}(\xi, M) \, \partial_t^{1/2} U(\xi, t)\Big|_{t=\tau_D(\xi,M)} . \tag{7.5}$$

If $\Delta\xi$ is not constant along the seismic line, each trace has to be weighted with a measure of the local horizontal trace increment in order to produce output amplitudes that are not falsified by the irregular geometry of the input data. This technique is usually part of what is called *acquisition footprint removal* in the literature, see, e. g., Canning and Gardner (1998). For 2.5D poststack migration, the calculation of a local $\Delta\xi$ is quite simple, see Figure 7.2. Each trace is weighted with half the distance of the two neighboring traces, i. e.,

$$\Delta\xi_i = \frac{\xi_{i+1} - \xi_{i-1}}{2} , \tag{7.6}$$

assuming that all traces in the poststack dataset are sorted with respect to ξ. For 2.5D common-offset prestack migration, we can use a similar weight. However, it has to be estimated for each trace in every offset bin separately, see Figure 7.3. Nevertheless, the estimation of the weighting factor corresponding to $\Delta\xi$ in equation (7.5) remains simple. Things become more tedious in 3D. Then, the discrete form of equation (5.2) reads

$$V(M) = \frac{-1}{2\pi} \sum \sum \Delta\xi_1 \, \Delta\xi_2 \, W_{DS}(\vec{\xi}, M) \left. \frac{\partial U(\vec{\xi}, t)}{\partial t} \right|_{t=\tau_D(\vec{\xi},M)} , \tag{7.7}$$

and each trace, in addition to the true-amplitude weight, has to be weighted with a local $\Delta\xi_1 \, \Delta\xi_2$, i. e., a local area that somehow describes the distribution of neighboring traces. For 3D poststack migration, we may use the Voronoi cell as an estimation for $\Delta\xi_1$ times $\Delta\xi_2$. What is a Voronoi cell? To explain its construction, we have to start off with a Delaunay[1] triangulation (Delaunay, 1934) of a point set that is defined by all trace positions of our seismic (poststack) dataset. The Delaunay triangulation of this point set in the plane is a collection of edges satisfying an empty circle property, i. e., for each edge we can find a circle containing the edge's endpoints but not containing any other points. It is in some sense the most natural way to triangulate a set of points. From the Delaunay triangles, we can construct the Voronoi[2] cells (Voronoj, 1908) in a simple way: we have to calculate the bisector of each previously estimated edge of the Delaunay triangulation. The cell that is defined

[1]The Russian mathematician Boris Delone was spelled in a phonetic transcription as "Delaunay" by his French publisher and, thus, his ideas became known under the pseudonym Delaunay.

[2]Actually, the Russian mathematician was called Voronoj but became know as Voronoi in the literature. Another mathematician, G. L. Dirichlet, studied similar problems as early as in 1850. Accordingly, the Voronoi diagram is sometimes also named Dirichlet tessellation.

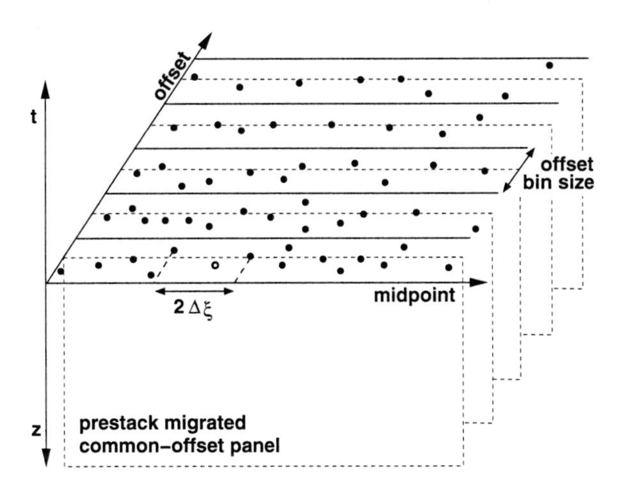

Figure 7.3: Calculation of a local trace weighting factor corresponding to $\Delta\xi$ in 2.5D common-offset prestack migration by means of two neighboring traces in an offset bin. The points denote trace positions, the circle marks the trace position for which the weighting factor is calculated.

by the bisectors and their intersection points around a given position (i. e., one point of our point set) is called Voronoi cell. The interior consists of all points in the plane which are closer to the particular lattice point under consideration than to any other, see also Figure 7.4. The Delaunay triangulation and the Voronoi diagram can be seen as dual (but conjugate) operations. The actual calculation of a Voronoi cell and its size might, of course, be implemented in a different way; the above description, however, might be the most intuitive one. For 3D prestack migration, we may extend the 3D poststack considerations shown here in a similar way as for the 2.5D situation mentioned earlier, see Jousset et al. (1999) for details and Jousset et al. (2000) for a real data example. Compared to the overall costs of migration, the estimation of Voronoi cells is relatively cheap and serves, thus, as a suitable tool to enhance the handling of amplitudes in the migration process which is particularly important in true-amplitude processing (see, e. g., Gesbert, 2002).

The true-amplitude weights in Kirchhoff migration depend on several parameters, as was shown in Chapter 5 and also by Spinner (2003) for measurement surfaces with topographic variations. For the kinematic part of the migration process, we need to consider only the elevation of the topography above a certain datum; for the handling of amplitudes, however, we also need an estimation of the local gradient of the measurement surface. For 2.5D migration, a simple 1D spline interpolation can be used to describe the topography in the direction of the seismic line. In this way, the local dip of the measurement surface is readily available from the spline coefficients. Unfortunately, in 3D one cannot use a 2D spline interpolation. The calculation of the spline coefficients would be too time-consuming. Moreover, most 2D spline interpolation implementations assume a regular distribution of lattice points, which is not the case in our situation and would require a regularization prior to the interpolation. We need to think of a different solution. As we use a Delaunay triangulation anyway (in order to estimate the factor with which to weight each trace in the summation process), it is obvious to use the triangulation also for the estimation of the local gradient of the measurement surface. If the elevation is taken into account, the triangles are no longer located in a plane but define a surface in 3D, see Figure 7.5. Now we can calculate a normal vector for each triangle and estimate the local

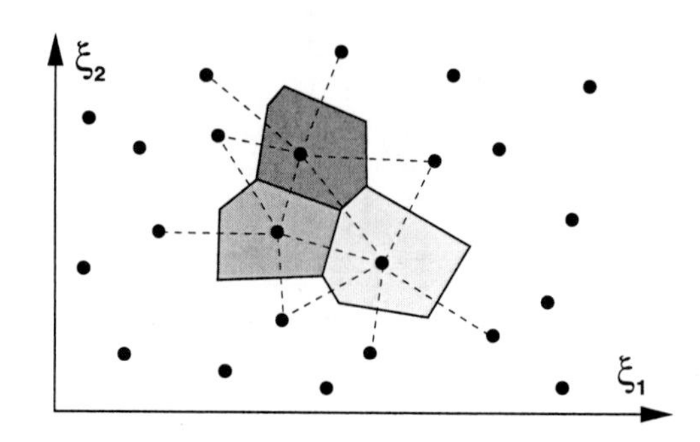

Figure 7.4: Voronoi cells. The points denote trace positions while the dashed lines denote some edges of a Delaunay triangulation. The shaded areas are the Voronoi cells used to weight each trace in a migration process.

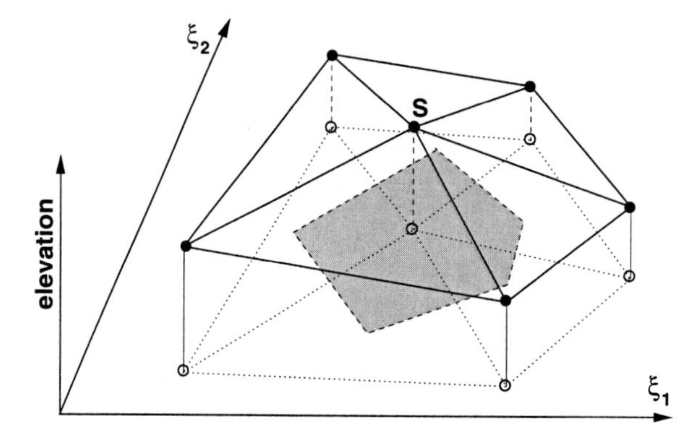

Figure 7.5: Calculation of the local gradient of the measurement surface by means of the Delaunay triangulation. Normal vectors for all relevant triangles are calculated and then subsequently combined and averaged to approximate the surface gradient at a lattice point S. The shaded area denotes the Voronoi cell.

gradient at a lattice point by properly weighting the normal vectors of adjacent triangles. This allows to calculate precise true-amplitude weighting factors while restricting the overhead of the necessary computations. An example will be shown in the next chapter.

7.6 Summary

In this chapter, several aspects of migration were discussed, some of them are general while others are specific for Kirchhoff migration. The last named method has a long tradition but is still frequently applied and turned out to be a workhorse in seismic data processing. Several reasons were mentioned; the most important one might be the accuracy and relative cheapness of Kirchhoff migration compared to other methods. In order to perform the migration process, we need a suitable macrovelocity model and estimates of the Green's functions—both aspects were discussed, mainly with respect to practical applications. A method to assess the correctness of the velocity model based on the flatness analysis of events in image gathers was presented. If, in addition, amplitude studies are carried out in the image gather, such a so-called AVO or AVA analysis extracts information from the prestack migrated data that might help to gain a better understanding of geological features in the subsurface. However, this is only possible if we have high-quality data where amplitudes were treated in a true-amplitude sense throughout the whole processing sequence. This implies that topography is properly considered in the migration process and that all aliasing effects due to the discrete nature of data acquisition and data processing are avoided as far as possible. These aspects were discussed in detail and some solutions were offered. In the next chapter, a synthetic data example is presented that incorporates all previously mentioned aspects of true-amplitude Kirchhoff migration.

Chapter 8

Synthetic data example

As part of this thesis, a true-amplitude Kirchhoff migration program has been developed. This program is called *Uni3D* in allusion to the underlying theory presented by Hubral et al. (1996) and Tygel et al. (1996). All aspects that were discussed in the previous chapters were incorporated in Uni3D, so the program is able to perform 2.5D prestack as well as 2.5D and 3D poststack depth migration[1] in a flexible way. This implies, for instance, the implementation of true-amplitude weights as discussed in Chapter 5, the limitation of the migration aperture and the estimation of taper lengths according to Chapter 6, or the handling of arbitrary smooth velocity models, irregular acquisition geometries (acquisition footprint removal), and topographic measurements surfaces as presented in Chapter 7. All migration results shown in this thesis are processed by means of the program Uni3D.

In order to test the applicability and the quality of (true-amplitude) migration and the additioanl aspects mentioned above, one must start off with a synthetic data example because only in this case there is a controlled environment where results by means of the program Uni3D can be compared to analytically calculated results or the model itself. Such a synthetic data example is presented in this chapter.

8.1 Model and forward-modeled data

The model used for the following tests is shown in Figure 8.1. It consists of several homogeneous blocks with P-wave velocities in the range 2 km/s $\leq v_P \leq$ 5 km/s. The S-wave velocities are set to the constant ratio $v_S = v_P/\sqrt{3}$. The density varies in the model which is not shown here. Note that the x- and z-axes are scaled differently; the topography and the structures in the subsurface are not as steep as they appear in this figure. Nevertheless, the surface elevation varies in a range of about 700 m. A multicoverage (prestack) and a zero-offset (poststack) dataset (PP-reflections, only) were modeled at the measurement surface by means of dynamic ray tracing. For the zero-offset example, a shot spacing of 10 m is used; in the multicoverage dataset, shot and receiver group spacings are 20 m and the maximum offset is 2,000 m, resulting in a total of 60,000 traces. A 20 Hz zero-phase Ricker wavelet is used as source pulse, and the time sampling interval of each trace is 4 ms. Some noise was added to the synthetic seismograms in order to make them more realistic. The modeled zero-offset

[1]The 3D prestack depth migration part has not been implemented. This kind of depth migration requires a huge amount of computation time and the implementation would have gone beyond the scope of this thesis.

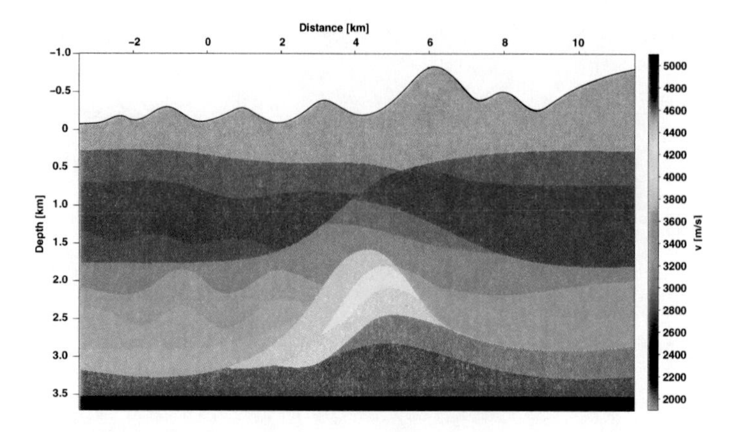

Figure 8.1: Blocky model with a curved measurement surface topography. The colors denote P-wave velocities which range from $v_P = 2,000$ m/s to $v_P = 5,000$ m/s. S-wave velocities are given by $v_S = v_P/\sqrt{3}$. The density varies in the model which is not shown here. Note the different scale of the horizontal axis compared to the vertical axis. The surface elevation varies in a range of about 700 m.

section is depicted in Figure 8.2 (left picture). The amplitudes of the events generally decrease with increasing traveltime mainly because of geometrical spreading effects. The late events can hardly be recognized and fade in the noise. The Green's function table was created in a smoothed version of the blocky model shown in Figure 8.1, see Grubb and Walden (1995) or Gold et al. (2000) for details on smoothing seismic velocity fields.

8.2 Pre- and poststack depth migration

If the modeled ZO section is migrated by means of Uni3D, we obtain the migrated image shown in Figure 8.2 (right picture). All reflectors are properly positioned which can be seen if one compares the migration result to the model shown in Figure 8.1. This means that a) the topography has been correctly handled in the migration process and b) the Green's functions were correctly estimated in the smooth macrovelocity model. This can be observed particularly with regard to the deep reflector at about 3.5 km depth that is perfectly flat. However, the whole image suffers from the background noise that is inherent in the input data. For poststack migration of real data, it is therefore of particular interest to simulate ZO sections with a high signal-to-noise (S/N) ratio. I will come back to this remark in the next chapter. The source pulse was correctly recovered because of the applied filter function in the diffraction stack. The isochron-like artifacts visible in the migrated image are due to the missing diffractions in the modeled data and due to (artifical) high amplitude anomalies that were caused by the ray-tracing program. This phenomenon was already explained in Chapter 6 on page 57. Such kind of artifacts can, however, only be eliminated manually if their cause is identified in the input data. Note that no artifacts due to the limited aperture are visible in the migrated image, compare Figure 6.8. This implies that the applied taper function was able to eliminate the unwanted effects.

The prestack depth migration image is shown in Figure 8.3 (left picture). To construct this figure, the individual common-offset gathers of the multicoverage dataset were migrated separately and, subsequently, summed up. The same Green's function table as in the poststack migration example was

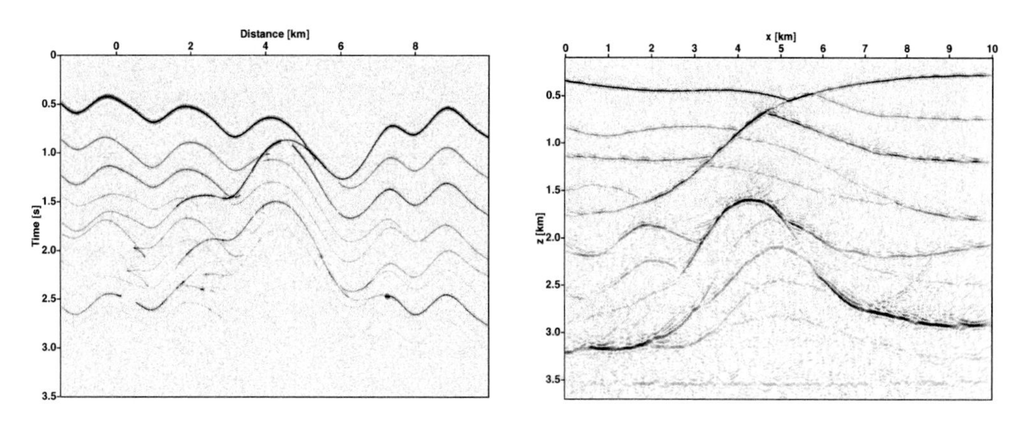

Figure 8.2: Left: calculated zero-offset section for the model shown in Figure 8.1. Some noise was added to the seismograms. Note that diffraction events were not modeled by the ray-tracing program. All reflection events are strongly influenced by the nonflat measurement surface. The amplitudes of the events generally decrease with increasing traveltime mainly because of geometrical spreading effects. Right: poststack migration result. The visible artifacts are either caused by the missing diffractions in the modeled data or high amplitude anomalies caused by the ray-tracing tool utilized to create the input data.

used. Once again, we observe that all reflectors are properly positioned—this can be easily checked by overlaying the actual reflector positions that were extracted from the model shown in Figure 8.1. The result is depicted in the right picture of Figure 8.3. Note that the isochron-like artifacts are due to the same reasons as in the poststack migration example. Such coherent artifacts remain in the migrated image even after summing up all offset panels whereas random noise is suppressed due to destructive interference. Thus, it can clearly be observed that the prestack migration results (a stack over individual common-offset migrated panels) is clearer than the poststack migration result (a single common-offset migrated panel, where the offset is equal to zero). One recognizes that the uppermost reflector is slightly blurred, especially at very shallow depths. This is an effect that can be explained by looking at an image gather. Two image gathers are displayed in Figure 8.4 for the positions $x = 2.6$ km and $x = 6.5$ km, respectively. They were extracted from the prestack depth migration result before a stack over all offset panels was applied. An offset-dependent stretch of the wavelets can be observed in the image gather. This phenomenon was explained in connection with the stretch factor m_D in Chapter 5. The stretch is usually all the more visible the shallower the reflector and the larger the offset. The reason for this behavior is the range of reflection angles that decreases with increasing depth and decreasing offset. Now, assume that all offset panels are summed up; if a significant stretch is present for an event in the image gather, the corresponding stacked result of the pulse will be blurred, even if the event is perfectly flat. This can be observed in Figure 8.3. As a consequence, image gathers are usually muted prior to the stack over all offsets as soon as the pulse stretch becomes larger than a predefined value. This ensures that the final prestack depth migrated image will be clear and will not lose resolution.

As can be observed in both image gathers in Figure 8.4, all events that define reflector positions for different offsets at the specified x-location are flat. This implies that the macrovelocity model used in the prestack migration process was data consistent and, thus, most probably represents the true velocity in the Earth. For real data, however, a verification can only be obtained by comparison with

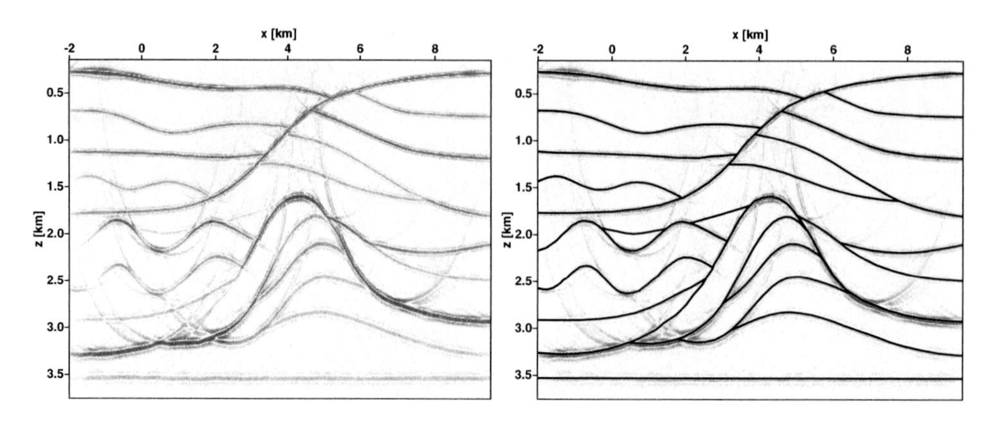

Figure 8.3: Left: prestack migration result. Note that the artifacts are either caused by the missing diffractions in the modeled data or high amplitude anomalies caused by the ray-tracing tool utilized to create the input data. Such "coherent" artifacts remain in the migrated image even after summing up all common-offset panels. Right: the same prestack depth migration image as in the left picture but with the true reflector positions (black lines) superimposed.

well data. In our case, a smoothed version of the true velocity distribution in the given model was used and, hence, it is not surprising that all image gathers show flat events.

8.3 AVO/AVA analysis

So far, we have only evaluated the kinematic properties of the poststack and prestack migration results by means of Uni3D. In order to evaluate the dynamic properties and, thus, the correctness of the true-amplitude weight function, amplitudes of the first reflector were picked in the above-mentioned image gathers (Figure 8.4). According to the theory of true-amplitude migration, each picked peak amplitude of the zero-phase wavelet in the depth domain should equal the PP reflection coefficient for the corresponding reflection angle (times an additional factor $\mathcal{A}$ that is usually assumed to be approximately 1, compare equation (1.1)). The extracted amplitudes define an AVO curve. However, to compare the picked amplitudes with analytically calculated values by means of Zoeppritz' equations, the AVO curve has to be transformed into an AVA curve. This is done by estimating the reflection angle of an incident P-wave for the first reflector at position x = 2.6 km and x = 6.5 km, respectively, for each offset and taking the topography into account. This is possible because the topography as well as the local dip of the reflector at the position of the image gather are known. As for the first reflector, we do not have to consider transmission losses (and the source strength is known and equal to 1), the normal incidence ($\alpha_{M_R} = 0$) plane-wave reflection coefficient should equal

$$R = \frac{v_2\rho_2 - v_1\rho_1}{v_2\rho_2 + v_1\rho_1} \,. \tag{8.1}$$

For the velocities and densities, we have to substitute the P-wave velocity and the medium density above and below the reflector at the relevant x-location, these being v_1 = 2 km/s, v_2 = 2.2 km/s, ρ_1 = 2 g/cm^3, and ρ_2 = 2.1 g/cm^3. Thus, we expect a normal incidence reflection coefficient of R = 0.072. As the velocity and density contrasts across the uppermost reflector are the same at both

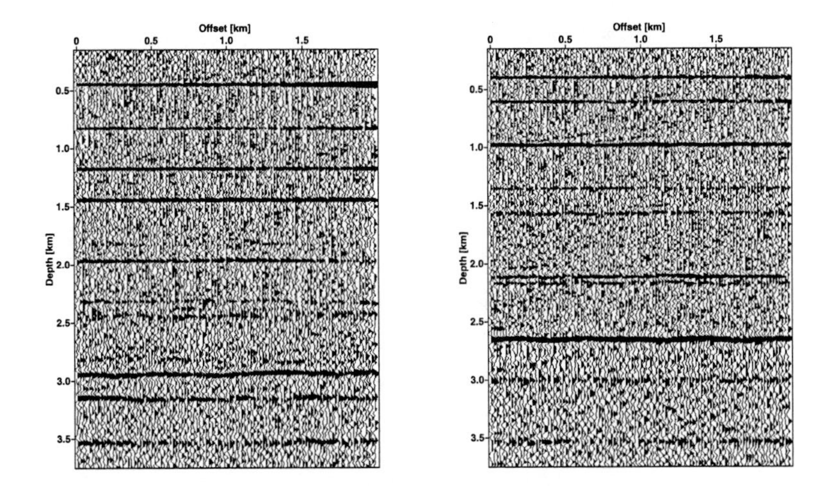

Figure 8.4: Two common-image gathers extracted at the positions $x = 2.6$ km and $x = 6.5$ km, respectively, from the prestack depth migration result before a stack over all common-offset panels was applied to produce Figure 8.3. All events in the image gather are flat which implies that the correct velocity model was used in the prestack migration process. Note the noise that is present in the image gathers. Random noise is suppressed when stacking over the offsets due to destructive interference.

locations of these two image gathers, we expect also the angle dependence of the output amplitudes to be the same. Due to the curved topography, the different depths of the considered image points, and the different dips of the reflector at these points, the maximum offset of 2,000 m corresponds to different maximum reflection angles for the two reflector points. The two AVA curves together with the curve analytically calculated by means of the Zoeppritz equation are displayed in Figure 8.5. The amplitudes extracted from the prestack depth migration result match the analytically calculated curve very well, except, of course, for some numerical errors and errors caused by the random noise in the data. The results show that the true-amplitude weight was able to correctly remove the geometrical spreading effect from the input data. In addition, it implies that the topography was correctly handled both kinematically and dynamically. Note that a similar study can be performed for all reflectors in the migrated image. However, for deeper reflectors the analysis is more cumbersome as transmission losses need to be considered in order to compare the numerical results with analytically derived values. As stated in Chapter 5, these additional amplitude effects do not pose any restrictions to the concept of true amplitudes: the true-amplitude migration process achieves its goal of correcting for geometrical spreading effects for all other reflectors like it did for the shallow reflector.

8.4 Irregular acquisition geometries

In Chapter 7 it was mentioned that we need to estimate a local $\Delta\xi$ (2.5D) or a local area $\Delta\xi_1$ times $\Delta\xi_2$ (3D) in order to correctly weight each trace in the migration process and, hence, to remove acquisition footprints. This is independent of the true-amplitude weight function applied. In simple migration programs, only an average trace distance is considered, resulting in unreliable amplitude information after migration. To test the effects of irregular acquisition geometries on migrated amplitudes, a

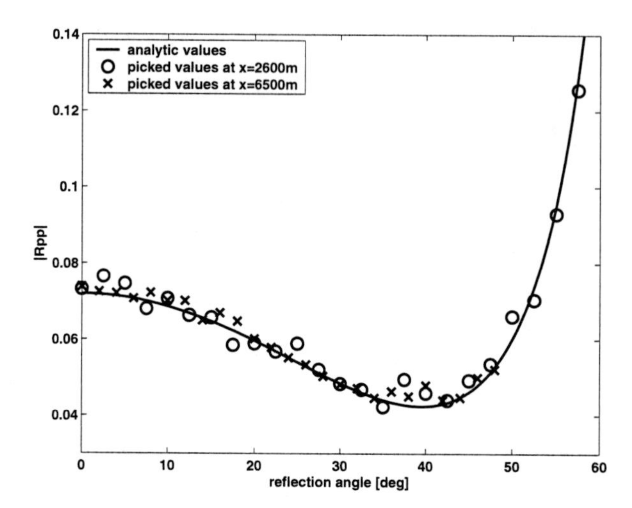

Figure 8.5: A comparison of extracted amplitudes of the shallow reflector after prestack depth migration with an analytically calculated AVA curve. The circles denote picked values at the location x = 2.6 km and the crosses values extracted at x = 6.5 km. The curves match very well.

special version of the synthetic dataset described on page 85 was created: the trace positions were randomly placed along the measurement surface in such a way that the average trace distance equals 12.5 m. Then, the newly simulated dataset was migrated twice: firstly by considering only an average trace distance of 12.5 m, secondly by considering the local trace distance that varies between 8 m and 17 m. Afterwards, the amplitudes along the first reflector in the migrated image for the offset equal to zero were picked; these amplitudes should reflect the normal incidence reflection coefficient. In this way, the test on irregular acquisition geometry is also a further test of the true-amplitude weight function. The results are shown in Figure 8.6. The left picture shows the picked values of the peak amplitude along the shallow reflector if an average value for the trace distance $\Delta\xi$ is used in the diffraction stack. Although the picked values (solid line) are in the magnitude of the correct normal incidence reflection coefficient (dashed line), we observe strong fluctuations which will hamper further amplitude analysis (this effect shows up in both the prestack and poststack case). The picture on the right-hand side of Figure 8.6 shows the same picked values extracted after a migration where the local trace spacing according to Chapter 7 was considered. The picked values match the normal incidence reflection coefficient almost perfectly without fluctuations. Note that even the sharp contrast that is caused by the different impedance contrast at the shallow reflector in the region between $x \approx 5$ km and $x \approx 6$ km (compare Figure 8.1; the normal incidence reflection coefficient for the first reflector is $R = 0.115$ in this region) is correctly recovered. It is easy to figure out that amplitude studies in the situation depicted in the picture on the right-hand side of Figure 8.6 are more reliable and simpler to perform than in the situation depicted in the picture on the left-hand side. In order to obtain high-quality amplitudes in migrated images, the local trace distance must obviously be taken into account.

To test poststack true-amplitude Kirchhoff migration of a 3D dataset with irregular geometry, I extended the model shown in Figure 8.1 in the y-direction. The subsurface structure and the model parameters are the same as for the 2.5D experiment, with identical extension into the third dimension. A 3D ZO section for randomly selected trace positions along the measurement surface with

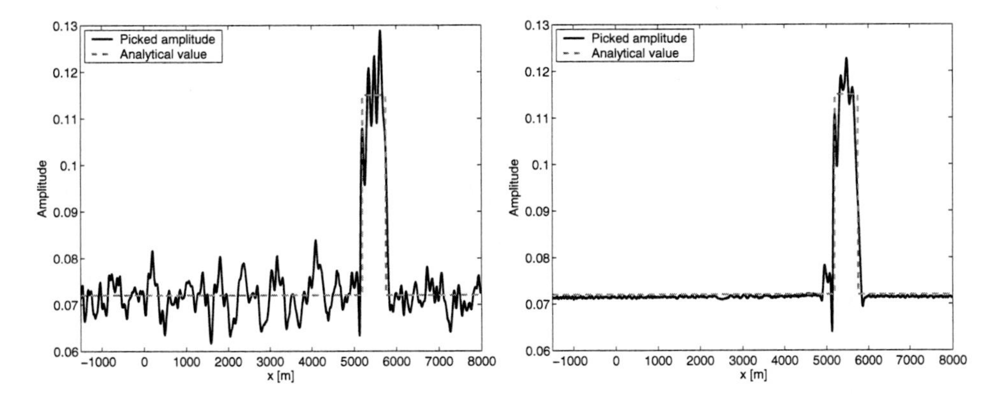

Figure 8.6: The effect of an irregular acquisition geometry on migration amplitudes. Left: picked amplitudes for offset equal to zero after 2.5D true-amplitude migration (solid line) where only an average trace distance was considered, and correct normal incidence reflection coefficient (dashed line). Strong fluctuations can be observed. Right: the same picked amplitudes after a 2.5D true-amplitude migration where the local trace distance was considered. We observe an almost perfect match of the picked amplitudes and the normal incidence reflection coefficient.

topographic variations was calculated by dynamic ray tracing. The trace positions have an average distance of 10 m, both in the ξ_1- and ξ_1-direction. As I did in the 2.5D case, this 3D section was then migrated in two different ways: firstly considering only the constant average trace spacings $\Delta\xi_1$ and $\Delta\xi_2$, and secondly considering the spacings for each trace locally by means of the Voronoi cell described in Chapter 7. Some Voronoi cells that were calculated for the irregularly distributed trace positions are shown Figure 8.7. Note that these cells usually cover the whole migration aperture in the plane $z = 0$.

Figure 8.8 (left picture) shows the picked peak amplitude (solid line) of the uppermost reflector after a 3D true-amplitude poststack migration where only average trace spacings were considered. Strong fluctuations can be observed that hamper further amplitude analyses. The picture on the right-hand side in Figure 8.8 shows the same picked peak amplitude after a 3D true-amplitude poststack migration with the randomly distributed input traces weighted in the diffraction stack by means of the area of the Voronoi cell. The fluctuations are almost eliminated and, thus, further amplitude analyses are much more reliable. In this sense, the Voronoi cell is a suitable instrument to improve the accuracy of true-amplitude Kirchhoff migration. AVO or AVA analyses after 3D prestack migration will, of course, also benefit from this approach. Data examples for such a prestack migration with emphasis on true amplitudes were presented by, e. g., Canning and Gardner (1998) or Gesbert (2002).

The data example in Figure 8.8 (right picture) also shows that the true-amplitude weight function was able to correctly remove the geometrical spreading effects from the input data during the diffraction stack. As expected, the picked amplitudes (solid line) follow the analytically calculated values (dashed line; $R = 0.072$ and $R = 0.115$, respectively) which implies that the topography was correctly handled in the migration. According to Chapter 7, the estimation of the surface dip was performed by weighting the normal vectors of adjacent Delaunay triangles. If one compares the 3D migration result (Figure 8.8) with the results of the 2.5D migration (Figure 8.6) one observes that the absolute values of the picked peak amplitude for the first reflector are slightly smaller in the 3D case. This is not sur-

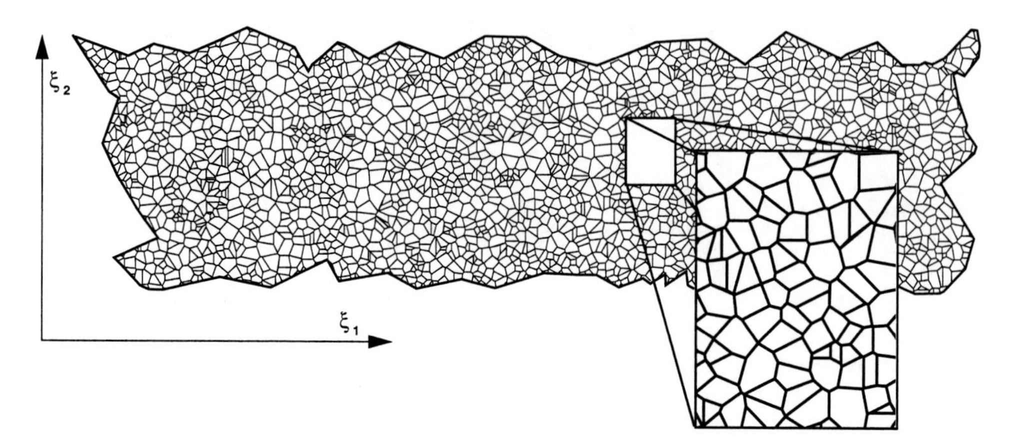

Figure 8.7: A part of the aperture in the plane $z = 0$ showing the Voronoi cells for the irregularly distributed trace positions of a 3D poststack section. In the zoomed area, one can clearly identify the cells and their different sizes with which traces are weighted in the 3D poststack migration process to remove acquisition footprints.

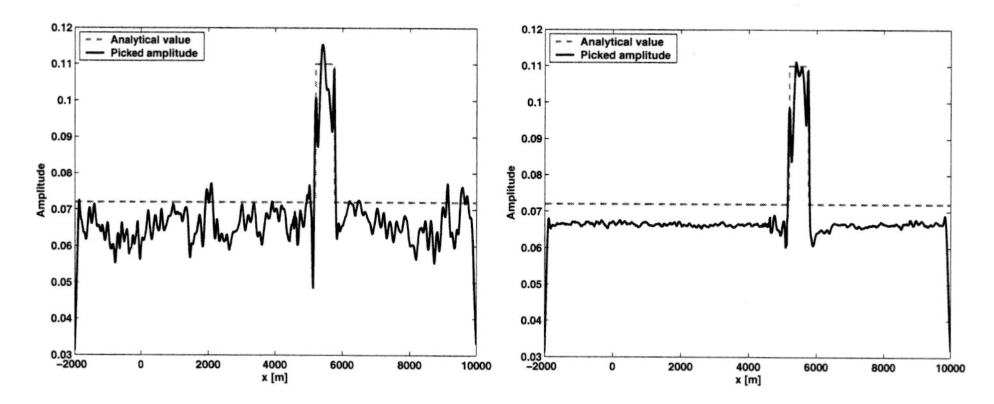

Figure 8.8: The effect of an irregular acquisition geometry on migration amplitudes. Left: picked amplitudes for offset equal to zero after 3D true-amplitude poststack migration (solid line) where only average trace spacings were considered, and correct normal incidence reflection coefficient (dashed line). Strong fluctuations can be observed. Right: the same picked amplitudes after a 3D true-amplitude poststack migration where the local trace weighting factor $\Delta\xi_1$ times $\Delta\xi_2$ was considered by means of Voronoi cells. We observe that fluctuations in the amplitude behavior are eliminated.

prising; the effect arises because the analytical approximation of the second integral in equation (5.2) by means of the stationary-phase method in 2.5D migration schemes (compare equation (5.20)) is usually more accurate than the numerical calculation by means of a stack used in 3D migration schemes (see also Chapter 5 and especially Figure 5.3). Note that a true-amplitude migration will achieve its goal of correcting for geometrical spreading effects for all other reflectors as well, independently of additional factors that affect the amplitude behavior.

8.5 Summary

In this chapter, several aspects of true-amplitude Kirchhoff migration were tested by means of a moderately complex model. Synthetic pre- and poststack datasets were modeled that formed the input for various migration tests which were accomplished by means of the program Uni3D. This program was developed as part of my thesis. The pre- and poststack depth migration results show that the kinematic aspects of the Kirchhoff migration process are correctly handled, i. e., reflectors are correctly positioned after migration. By analyzing common-image gathers, AVA curves were compared to analytically calculated angle-dependent reflection coefficients. It turned out that the true-amplitude weighting functions were able to achieve the goal of removing geometrical spreading effects in the diffraction stack migration. Topographic variations of the measurement surface are treated by spline interpolation (2.5D) or Delaunay triangulation (3D); these approaches proved to be reasonable solutions. Furthermore, it was shown that the irregular trace increments which we encounter when processing data acquired with irregular acquisition geometry play an important role if precise amplitude information is desired after migration. This problem is usually dealt with under the term acquisition footprint in the literature. A local trace increment $\Delta\xi$ is used in the 2.5D migration scheme, while the 3D scheme utilizes the Voronoi cell to estimate the trace weighting factor $\Delta\xi_1$ times $\Delta\xi_2$. In this way, distortions and fluctuations of amplitudes in the migrated images are minimized. As a consequence, amplitude analyses after migration become more reliable.

In the next chapter, true-amplitude Kirchhoff migration and the implemented program Uni3D are put into the context of a data-driven imaging workflow based on the Common-Reflection-Surface stack. This workflow is demonstrated on a real data example.

Chapter 9

A seismic reflection imaging workflow based on the CRS stack

The data-driven Common-Reflection-Surface (CRS) stack method is a generalized multi-dimensional and multi-parameter stacking velocity analysis tool. In its application, emphasis has so far mainly been put on its ability to produce simulated zero-offset sections of high S/N ratio. However, the method also yields a number of so-called kinematic wavefield attributes. With these attributes, an entire seismic reflection imaging workflow can be defined. Here, I briefly present the involved methods and demonstrate their usage and some of the possibilities in combination with a Kirchhoff depth migration on a real data example. In other words, I start with the preprocessed multicoverage data in the time domain and end up with an image in the depth domain that can be used for further lithostratigraphic analyses.

9.1 Introduction

As we have seen in previous chapters, seismic reflection data processing aims at obtaining the best possible image of the subsurface, either in the time or in the depth domain. Especially in regions with complex geological structures or for data with low signal-to-noise (S/N) ratio, this is a demanding task that usually requires extensive human interaction. One possible alternative is to automatically extract as much information as possible directly from the measured data. The continuous advances in computing facilities make such data-driven approaches (e. g., Hubral, 1999) feasible which, thus, have increasingly gained in relevance in recent years. Here, I focus on one of these methods, the CRS stack (e. g., Müller, 1999; Jäger et al., 2001; Mann, 2002), and its integration into the seismic reflection imaging workflow (especially with respect to the depth migration part) as is shown in a simplified way in Figure 9.1. A general overview of the main data processing steps has already been given in Chapter 1 (Figure 1.4).

As is shown below, the CRS stack provides a simulated zero-offset (ZO) section of high S/N ratio and is, thus, a superior substitute for the conventional NMO/DMO/stack approach (e. g., Yilmaz, 2001). Besides the improved ZO simulation, there is an additional benefit that is obtained with the CRS stack: instead of the usual stacking velocity, the process yields an entire set of so-called kinematic wavefield attributes. This additional information is very useful in further processing. Firstly, the

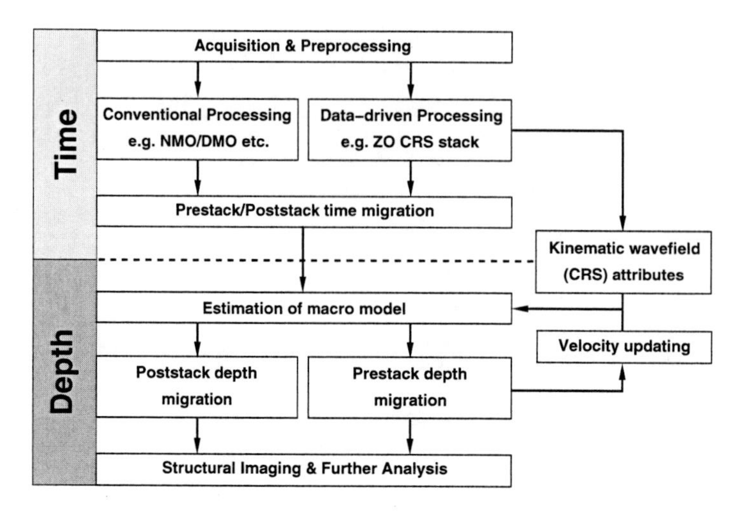

Figure 9.1: Integration of the Common-Reflection-Surface stack into the seismic reflection imaging workflow.

attributes can be utilized in the determination of a velocity model: an attribute-based generalized Dix-type inversion scheme for layered models has been discussed by Majer (2000) and Biloti et al. (2002), whereas a tomographic inversion has recently been introduced by Duveneck and Hubral (2002), see also Duveneck (2004). The latter yields a smooth velocity model well suited for ray-based depth imaging, e. g., by means of the migration algorithm presented in Chapter 5. In contrast to conventional inversion methods, the tomographic approach discussed here does not assume continuous reflection events in the data and requires only minimum picking effort. As was pointed out in Chapter 7, the quality of the (initial) macrovelocity model is crucial to successful depth imaging; the closer the model to the true effective velocity in the subsurface, the shorter the turn-around time. Secondly, properties like, e. g., the geometrical spreading factor (Vieth, 2001) or the projected Fresnel zone (Mann, 2002) can be estimated by means of the kinematic wavefield attributes, or they help to distinguish between reflection and diffraction events (Mann, 2002). Finally, they can be utilized in combination with the determined velocity model and the simulated ZO section in the Kirchhoff migration process itself to determine an optimal migration aperture. This technique, however, has not been implemented so far and needs further investigations.

In the approach presented here, flexible pre- and poststack processing strategies are available. They are based on combinations of conventional, model-based technologies and emerging data-driven imaging methods. Starting with the preprocessed multicoverage data in the time domain, the CRS stack yields sufficient information for transforming these data into an image in the depth domain. I demonstrate the potential of this approach on a real data example.

9.2 The seismic reflection experiment

The aim of the seismic reflection experiment that was carried out in the Oberrheingraben near Karlsruhe can be formulated as follows: it should provide a detailed structural subsurface image up to a depth of about 3 km in order to determine the location of key reflectors and fault zones. Besides of this

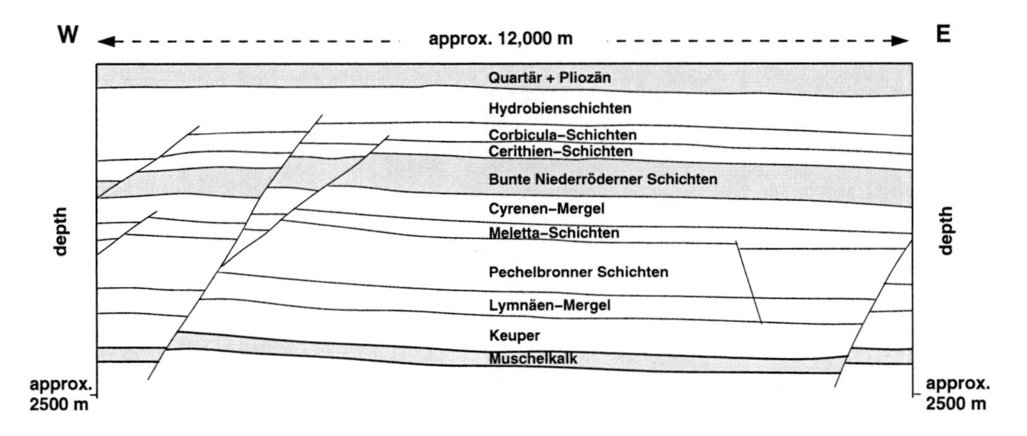

Figure 9.2: Geological profile along the direction of the seismic lines. The seismic experiment aimed at improving the structural image of the subsurface and the accuracy in position and dip of the already known reflectors.

primary goal, further analyses like, e. g., AVO or AVA, should provide additional information about the geological features and subsurface properties in the target zone. A geological profile is shown in Figure 9.2; as can be observed, only large-scale features were known prior to the current experiment, mainly from reflection seismic experiments that took place in the 1970's. Unfortunately, the acquisition technology at that time was not as sophisticated as it is today and, in addition, only depths up to 1500 m were explored in detail.

In order to meet the guidelines, two 2D seismic surveys (denoted by A and B in the following) were conducted with an average length of the seismic lines of about 12 km. The measurements were accomplished by *Deutsche Montan Technologie GmbH*, Essen by order of *HotRock EWK Offenbach/Pfalz GmbH*, Karlsruhe. Approximately 3,000 geophones were laid out with a receiver group spacing of 50 m (fixed spread) and 12 geophones per group. Three vibrators made up the seismic source, one of them is shown in Figure 9.3. The seismic signal used in the experiment was a linear upsweep (12 Hz - 100 Hz) of 10 s duration. The distance between source positions was 50 m, and six vibrator signals were (vertically) stacked for each source point. At each geophone, the reflected signals were recorded with a 2 ms time sampling interval for about 4 s.

The conventional preprocessing of the recorded data included geometry setup (crooked line), geometrical spreading correction, trace editing, field static correction (determined from refraction seismics), amplitude correction, minimum-phase shaping, bandpass filtering, muting, surface-consistent deconvolution, and static correction. Some of these steps, however, strongly affect the amplitudes of the recorded signals and, thus, do not allow depth imaging with true amplitudes. Hence, all results presented in this chapter are only reliable from a kinematic point of view. The dynamic part, i. e., the amplitudes, in migrated images can merely be interpreted in a qualitative, but not in a quantitative way. In the future, further investigations on data where some of the intermediate preprocessing steps are omitted in order to retain the amplitude information as far as possible will yield additional information on (litho)stratigraphic properties of the subsurface.

In the following, I briefly explain the steps that were applied to image the preprocessed data. The underlying basic principles are presented and the results of each processing step are shown.

Figure 9.3: One of three vibrator trucks that were used as sources in the seismic reflection experiment. An upsweep (approx. 12 Hz to 100 Hz) was transmitted to the Earth via the steal plate at the bottom of the truck.

9.3 The Common-Reflection-Surface (CRS) stack

The simulation of stacked ZO sections is routinely applied to enhance the S/N ratio and reduce the amount of seismic data for further processing. A conventional approach to achieve this goal is the application of normal-moveout (NMO) and dip-moveout (DMO) corrections to the multicoverage dataset followed by a subsequent stack along the offset axis, usually denoted as NMO/DMO/stack. The CRS stack is a powerful alternative to this conventional approach that can be seen as a generalized multi-dimensional high-density stacking-velocity analysis tool. It produces a simulated ZO section from the multicoverage data in a purely data-driven way. In addition, the CRS method provides a number of kinematic wavefield attributes associated with each ZO sample to be simulated. These attributes locally describe the shape of reflection events in the data. In the 2D case, the CRS stack fits entire stacking *surfaces* to the events rather than only stacking *trajectories*, as is done in conventional ZO simulation methods. Thus, far more traces contribute to each ZO sample, resulting in a higher S/N ratio, even for data of poor quality. As was pointed out in the last chapter, a ZO section with such a high S/N is desired in order to a obtain clear image of the subsurface after poststack migration.

The 2D stacking operator for a ZO sample (t_0, x_0) reads

$$t^2\,(x_m, h) = \left[t_0 + \frac{2\,\sin\alpha\,(x_m - x_0)}{v_0}\right]^2 + \frac{2\,t_0\,\cos^2\alpha}{v_0}\left[\frac{(x_m - x_0)^2}{R_\mathrm{N}} + \frac{h^2}{R_\mathrm{NIP}}\right], \qquad (9.1)$$

where the half-offset h and the midpoint x_m between source and receiver describe the acquisition geometry and v_0 is the near-surface velocity assumed to be locally constant. The remaining three parameters are the kinematic wavefield attributes. They describe the propagation direction (α) and radii of wavefront curvature (R_NIP, R_N) associated with two hypothetical experiments observed at $(z = 0, x_m)$. The NIP (normal incidence point) wave is the hypothetical wave that would be obtained

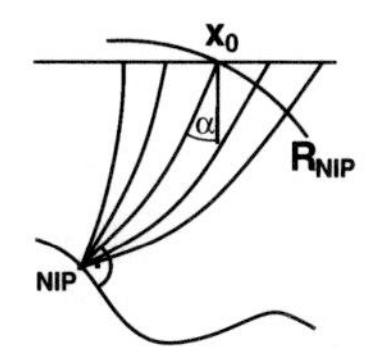

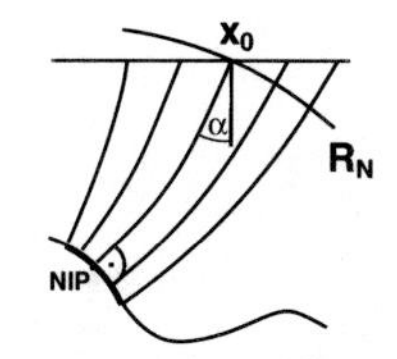

Figure 9.4: Hypothetical NIP wave (left) and normal wave (right) experiments. The angle α describes the emergence direction of the two hypothetical waves at the surface location x_0. The parameters R_{NIP} and R_{N} are the observed radii of wavefront curvature associated with these waves at x_0.

by placing a point source at the NIP of the ZO ray. The N (normal) wave is the hypothetical wave that would be obtained by placing a small exploding reflector element at the NIP of the ZO ray. These hypothetical experiments are illustrated in Figure 9.4.

To determine the attributes of the CRS operator fitting best an actual reflection event, a coherence analysis is performed along stacking operators in the multicoverage data with different sets of kinematic wavefield attributes. The best fitting operator yields the highest coherence. This analysis is repeated for each ZO sample to be simulated, irrespective of whether there is an actual reflection event. In case of conflicting dip situations, also local coherence maxima have to be considered. Based on coherence analysis, the entire CRS approach can be applied in a noninteractive way and without the need for any a priori knowledge of a macrovelocity model.

The CRS stack was applied to the multicoverage datasets mentioned in Section 9.2. The determination of the optimum stacking operators and their associated wavefield attributes has been performed in separate steps with one search parameter each. This pragmatic approach of using certain subsets of the prestack data was introduced by Müller (1998). As this approach fails for a few ZO locations, a smoothing algorithm has been applied to the attribute sections which uses a combination of a median filter and averaging. Coherence and local dip of the reflection events are taken into account during smoothing. From a theoretical point of view, the smoothing is justified as the wavefield attributes can only vary smoothly along the reflection event and are virtually constant along the seismic wavelet (Mann and Höcht, 2003). The smoothed attribute sections have served as input to a local three-parameter optimization using the full spatial stacking operator (9.1) and the entire prestack data which yielded the final wavefield attributes for stacking.

The simulated ZO sections are displayed in Figure 9.5, both for seismic line A and B. A maximum offset of 2,000 m in the input data was considered for the simulation of each ZO sample in the output. The stacking aperture takes the projected ZO Fresnel zone into account which has been estimated from the wavefield attributes.

The associated coherence and kinematic wavefield attribute sections are depicted in Figure 9.6 for the seismic line B. Similar sections (not shown here) are obtained for the seismic line A. The coherence section (d) shows the maximum semblance value obtained along the CRS operator for each simulated ZO sample. At the location of actual reflection events, its value is influenced by the signal strength relative to the random noise along the reflection event, by the number of contributing traces, and by the fit of the CRS operator to the actual reflection event. All CRS attribute sections only have meaningful values where the coherence value is sufficiently high, i. e., on actual reflection events. Thus, unreliable parts of the attribute sections are masked out and shown in black. Picture (a) shows the emergence angle section; the emergence angle varies roughly between -20° and +20°, but the majority of ZO rays emerge nearly vertically at the measurement surface. The radius of curvature of the NIP wavefront

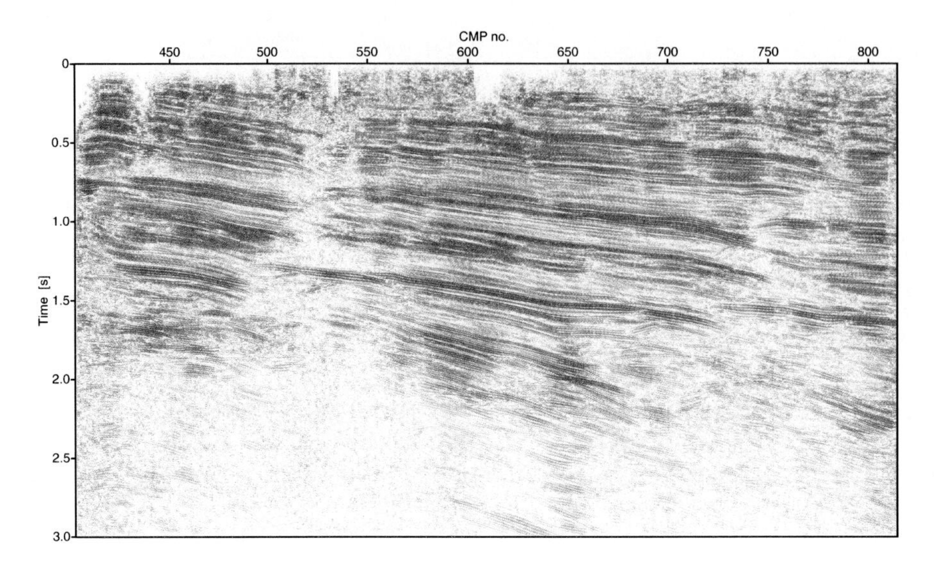

(a) Seismic line A

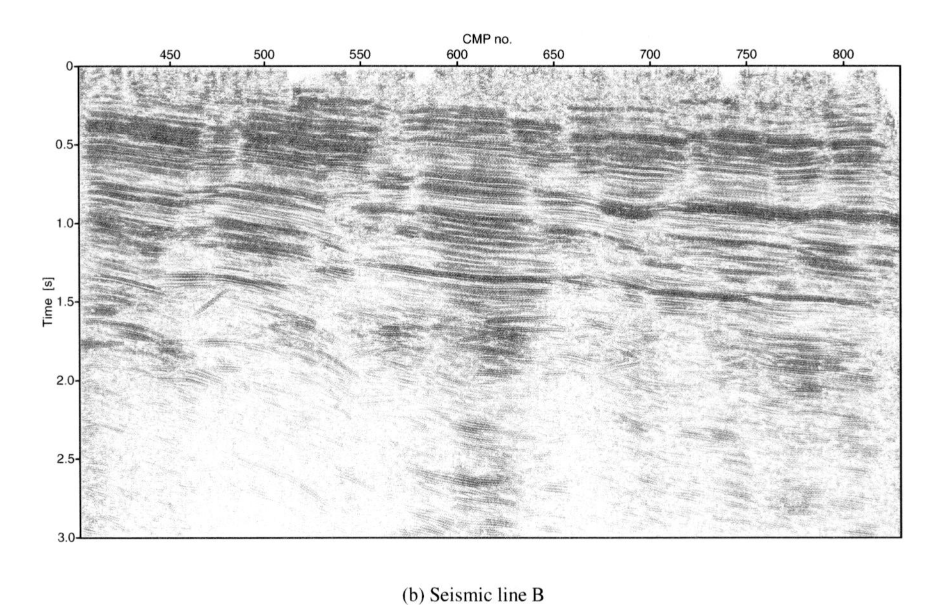

(b) Seismic line B

Figure 9.5: CRS-stacked ZO sections simulated from the multicoverage preprocessed datasets. The horizontal axis denotes the midpoint number. The time axis points downwards. ZO traveltimes up to 3 s were simulated. Note the high S/N ratio.

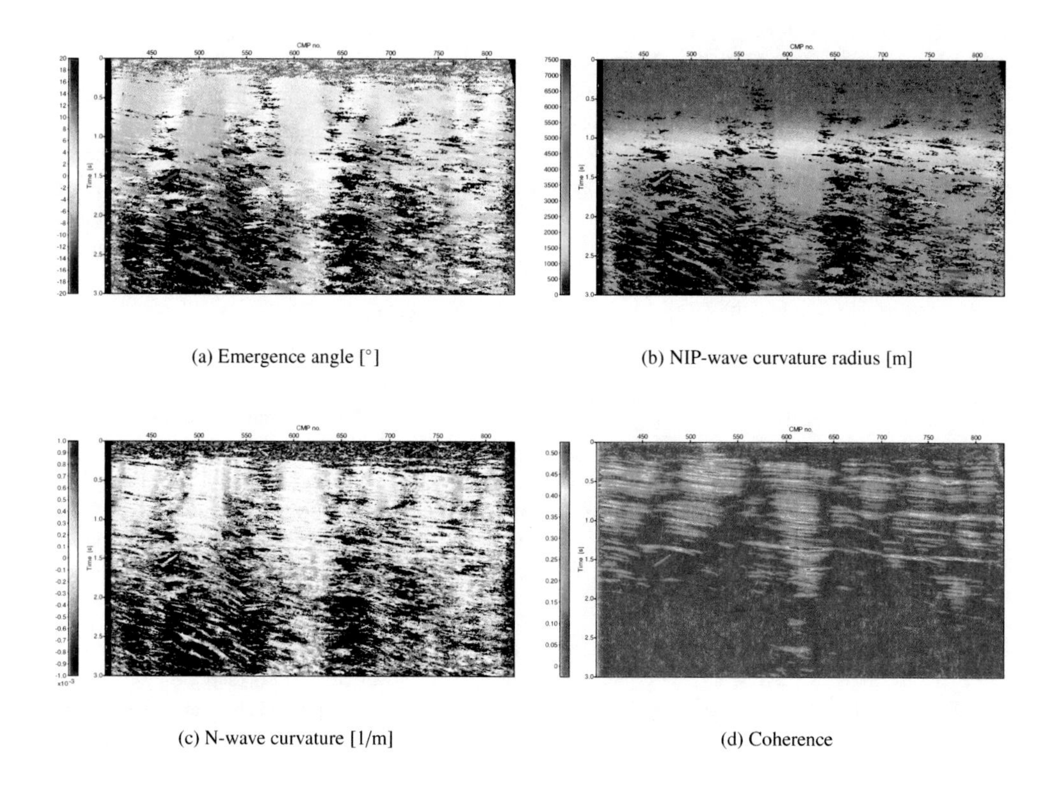

(a) Emergence angle [°]

(b) NIP-wave curvature radius [m]

(c) N-wave curvature [1/m]

(d) Coherence

Figure 9.6: Coherence section (d) and kinematic wavefield attribute sections (a)-(c) obtained by applying the CRS stack method to the multicoverage data of seismic line B. Unreliable parts of the attribute sections due to low coherence are masked out and shown in black. Note that the radius of curvature R_{N} is displayed as its inverse, i. e., the curvature of the normal wavefront.

is displayed in picture (b). In a constant velocity medium, this radius of curvature would coincide with the length of the normal ray (i. e., the distance to the NIP). Picture (c) shows the curvature of the normal wave.

Other real data examples of the CRS stack application, both for 2D and 3D industrial seismic data, can, e. g., be found in Trappe et al. (2001), Bergler et al. (2002), Cristini et al. (2002), or Trappe et al. (2003).

9.4 Velocity model determination

The estimation of a velocity model is one of the crucial steps in seismic depth imaging. As explained in Chapter 7, the model is often constructed iteratively, starting with an initial model and updating it by repeated prestack migration and analysis of residual moveouts in common-image gathers (CIGs). This is an expensive and time-consuming process. Approaches based on reflection tomography have the additional drawback of requiring extensive and often difficult picking in the prestack data.

The CRS technique offers an alternative which overcomes some of the drawbacks of conventional methods; the attributes R_{NIP} and α related to the NIP wave (Figure 9.4) at a given ZO location (x_0, t_0) describe the approximate multi-offset reflection response of a common reflection point (CRP) in the subsurface. Therefore, the NIP wave focuses at zero traveltime at the NIP if propagated into the subsurface in a correct model. This principle can be utilized in an inversion that uses the attributes R_{NIP} and α picked in the CRS-stacked section to obtain a laterally inhomogeneous velocity model for depth imaging. The CRS-stack-based velocity determination approach is realized as a tomographic inversion, in which the misfit between picked and forward-modeled attributes is iteratively minimized in the least-squares sense. The velocity model is defined by B-splines, i. e., a smooth model without discontinuities is used, which is well suited for ray-tracing applications.

As the attributes associated with each ZO sample already represent the multi-offset reflection response of a CRP, picking only has to be performed in the CRS-stacked ZO section. Due to the increased S/N ratio of the stacked section, the picking procedure is further simplified and may be automated based on the coherence section (Figure 9.6(d)). Because of the smooth model description, pick locations do not need to follow reflection events over consecutive traces. The approximate description of the multi-offset CRP response with CRS attributes, however, leads to a limitation of the allowed degree of lateral inhomogeneity in the model. Furthermore, a smooth model description may be inappropriate in some cases (e. g., salt bodies of complicated shape). Details of the method are described in Duveneck (2004).

In the data example, 793 points (seismic line A) and 1110 points (seismic line B), respectively, have automatically been picked in the the coherence section, Figure 9.6(d). The corresponding attributes have been simultaneously extracted from the R_{NIP} and α sections, Figures 9.6(b) and 9.6(a). These data entered into the inversion algorithm. The inversion result consists of the reconstructed velocity model and the reconstructed normal rays associated with the picks. The smooth velocity model $v(x, z)$ which is composed of B-splines with 21 nodes in x-direction and 16 nodes in z-direction is depicted in Figure 9.7. It should be mentioned that picking of multiple reflections in the data will, of course, affect the output velocity model and must be avoided. Therefore, manual checking of the automatic picks and possibly discarding some of them is mandatory.

9.5 Depth migration

For both seismic lines, pre- and poststack Kirchhoff depth migrations have been performed by means of the program Uni3D mentioned in Chapter 8. The Green's function tables were calculated in the reconstructed velocity models shown in Figure 9.7. A depth-dependent operator aperture was used in the Kirchhoff migration with a minimum value of 400 m at the top of the migration target zone and a maximum value of 2,000 m at the bottom. Linear interpolation of the operator aperture was applied in between. In this way, both the computation time and the aliasing were decreased without losing information or applying rigorous anti-aliasing filters that decrease the frequency content of the data and, thus, the possible resolution in the output domain. In the prestack migration process, offsets up to 2,000 m were considered. The spatial increments in the output were chosen Δx = 10 m and Δz = 5 m in order to avoid image aliasing and steps in the image of the slightly dipping reflectors. The poststack depth migration results are depicted in Figure 9.8.

To obtain the prestack depth migration images shown in Figure 9.9, the input multicoverage dataset was sorted into CO sections. Each CO section was then migrated separately. Afterwards, an offset- and

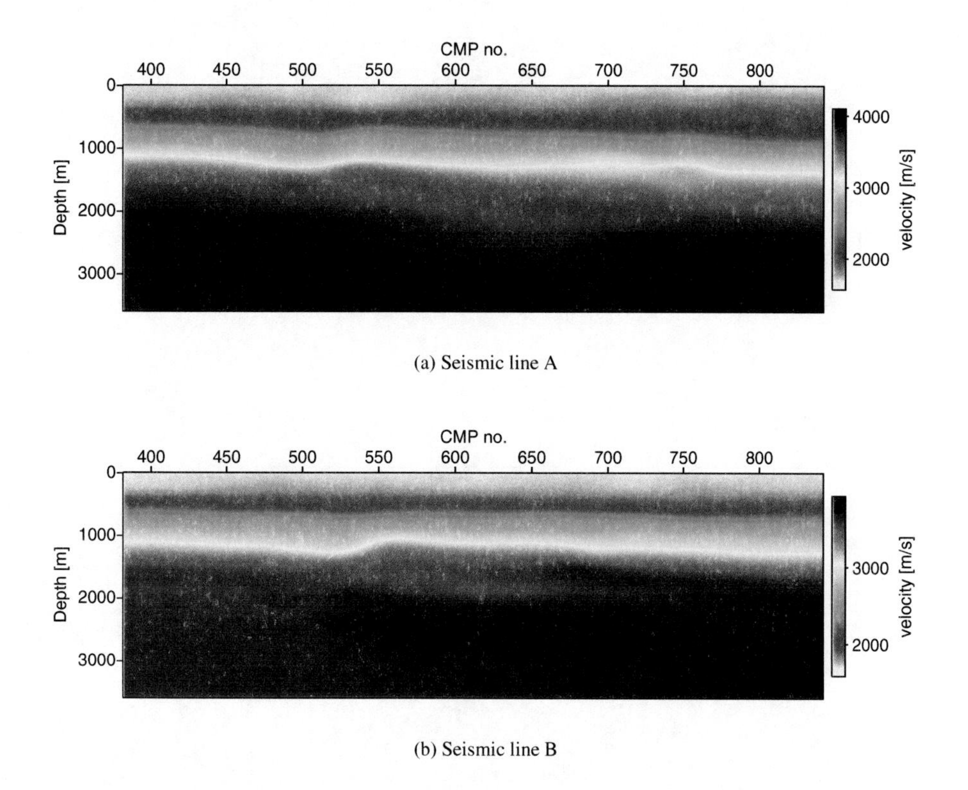

(a) Seismic line A

(b) Seismic line B

Figure 9.7: Smooth velocity models $v(x,z)$ obtained by a tomographic inversion approach based on kinematic wavefield attributes determined by means of the CRS stack method.

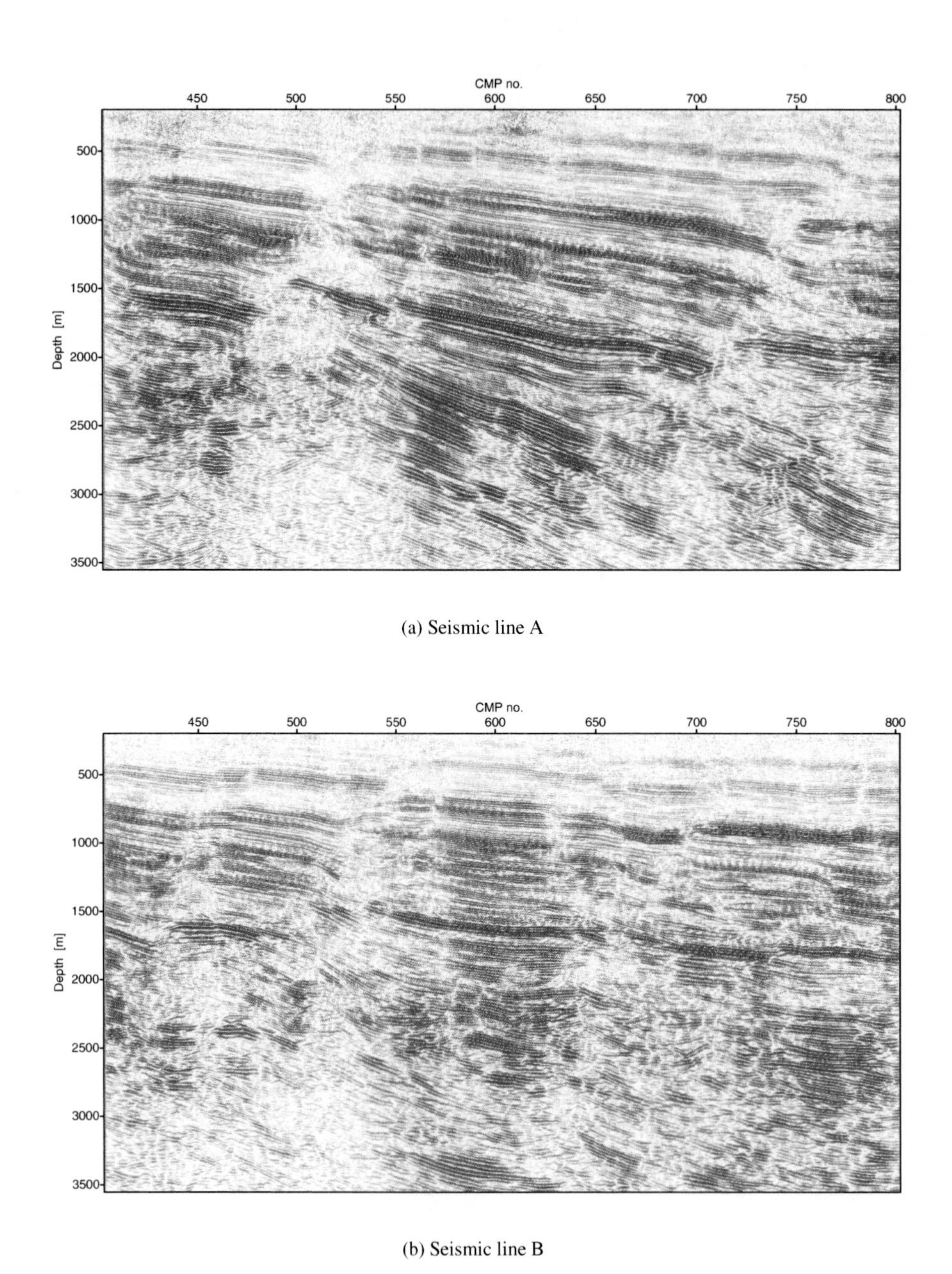

(a) Seismic line A

(b) Seismic line B

Figure 9.8: Poststack depth migration results obtained with the velocity models shown in Figure 9.7. The CRS-stacked ZO sections presented in Figure 9.5 were used as input.

depth-dependent mute was applied in the CIGs to avoid the stacking of stretched wavelets and, hence, a blurred (or completely destroyed) image. This procedure was explained in detail in Chapter 8. Some CIGs for seismic line A are shown in Figure 9.10. The mute function can easily be observed. Note that no velocity update was performed; nevertheless, most of the events in the CIGs are flat; strongly curved events mainly correspond to multiple reflections. In this data example, the CIGs indicate that the tomographic inversion based on CRS attributes was able to estimate the initial macrovelocity model very well.

The post- and prestack depth migration results show many structural details that complement the geological map (Figure 9.2) which has been available so far. In particular, many faults can readily be observed whose number exceeds all expectations. Vertical offsets of reflectors, deflection of reflectors, changes of reflector characteristics across faults, and fracturing are directly observable in the migrated sections. Well data that are available from an old nearby borehole show good agreement to key reflector positions in the migrated images. An initial interpretation of the prestack migrated images is shown in Figure 9.11. This preliminary interpretation is courtesy of N. Harthill, HotRock EWK Offenbach/Pfalz GmbH; it was made to determine structure and faulting and is only a small fraction of what may be accomplished by a quantitative interpretation of, e. g., reflector characteristics. Further geological and (litho)stratigraphic interpretations will follow. From the interpreter's point of view (N. Harthill, pers. comm.), the results of the CRS-stack-based imaging workflow have some major advantages compared to the results of the conventional seismic data processing that are also available:[1]

- In general, reflectors are imaged much better.
- Lateral variations in reflector characteristics can easily be observed.
- Reflectors beneath the Pechelbronner Schichten reflector (see Figure 9.2) are clearly imaged or imaged at all.
- Faults above *and* below the Pechelbronner Schichten reflector are clearly imaged.
- Faults may be traced from near-surface up to a depth of about 3,000 m.

For this data examples, the CRS-stack-based imaging workflow demonstrated its potential to provide all information required to successfully transform prestack and poststack data in the time domain into a structural image in the depth domain. The approach was applied in a highly automated manner with minimum human interaction. The convincing results, obtained both from seismic data processing and initial interpretation, show only some of the potential of the method. There will be further improvements, especially if a true-amplitude depth migration can be applied to the carefully preprocessed multicoverage data.

9.6 Summary

I have demonstrated that the CRS stack and the associated kinematic wavefield attributes can be used in seismic imaging applications which go far beyond the purposes for which the method was originally designed—the simulation of ZO sections with significantly improved S/N ratio. The kinematic

[1] Note that this is only a qualitative, i. e. subjective, statement.

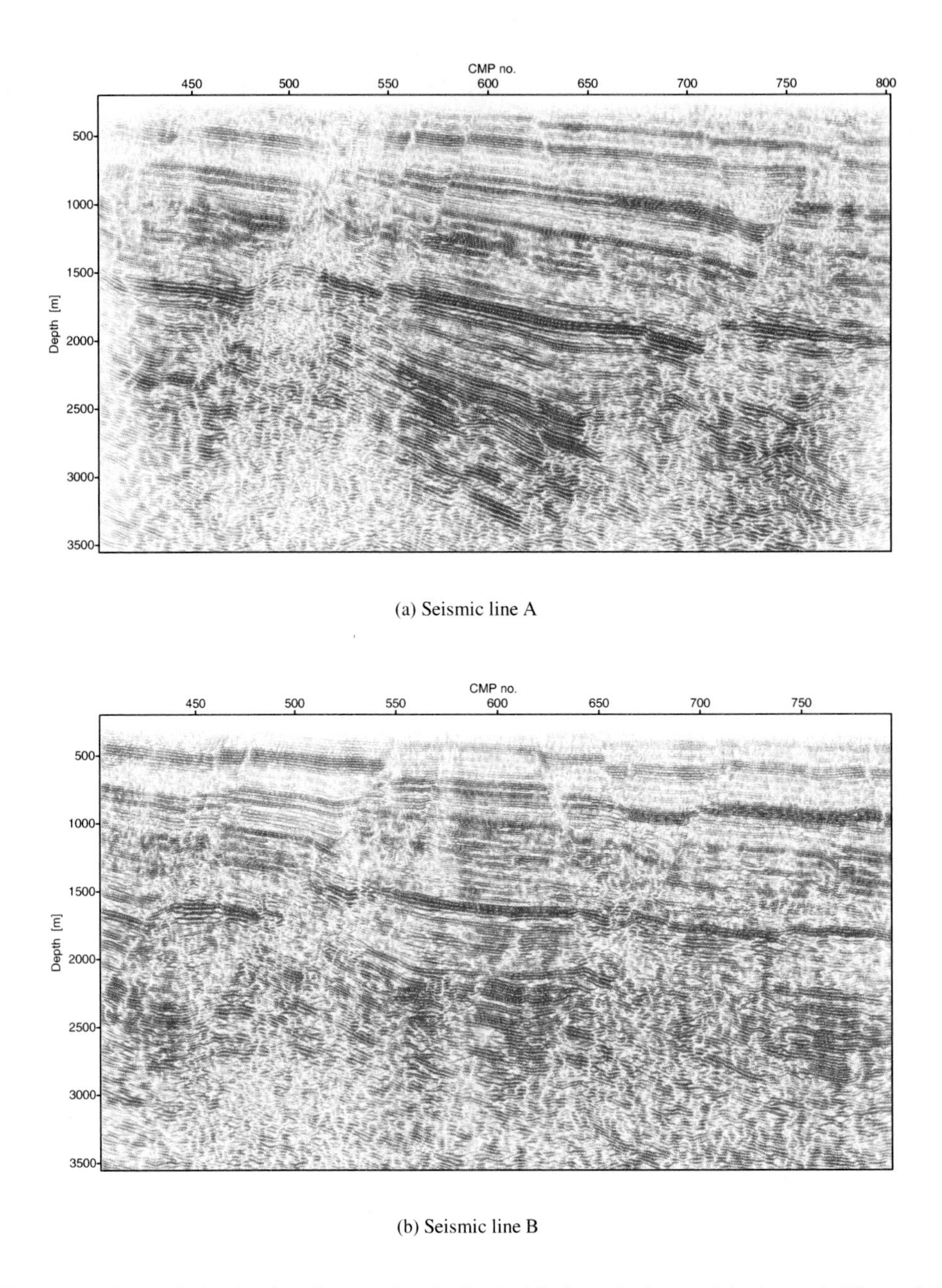

(a) Seismic line A

(b) Seismic line B

Figure 9.9: Prestack depth migration results obtained with the velocity models shown in Figure 9.7. After the migration of the individual CO sections, a mute was applied in the CIGs. Finally, all migration results for the individual offset bins were summed up along the offset axes to generate the pictures shown here.

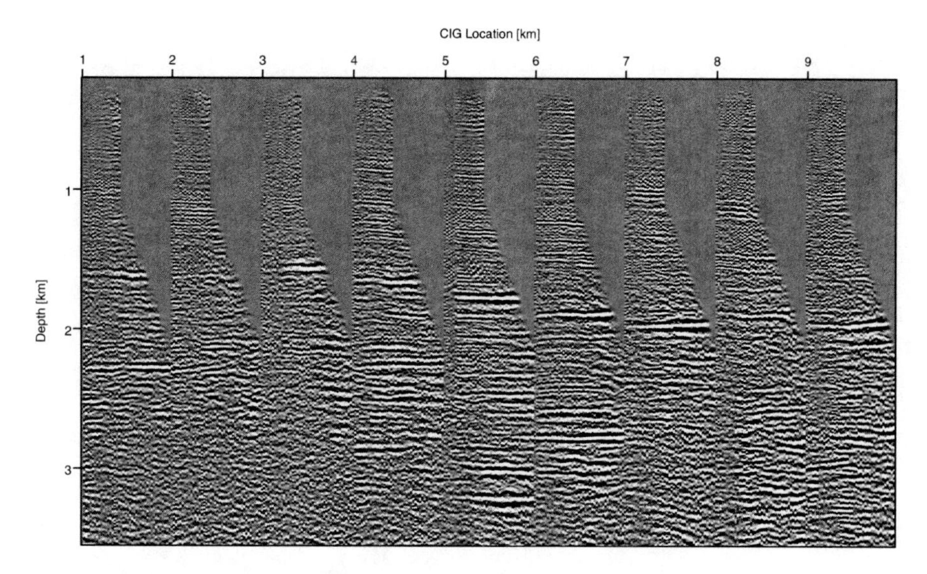

Figure 9.10: Some common-image gathers for several positions (denoted in km) along the seismic line A. The applied mute function can easily be observed. Note that most events are flat although no migration-based iterative velocity update was applied.

wavefield attributes contain information that can be used for the estimation of migration velocity models. The subsequent depth migration process profits from both, the CRS stack and the tomographic velocity estimation, as was clearly outlined in this chapter. Following this approach, imaging can be performed with a variety of case-specific strategies in a highly automated way. In particular, data of poor quality, land data suffering from topography and near-surface effects, or data with irregular acquisition geometries are expected to benefit from this approach.

Finally, here are some comments on the run time estimates related to this seismic reflection imaging approach; all estimates are based on a system consisting of a single CPU (i386 architecture) with approx. 1.6 GHz and 1 GB RAM:

- the CRS stack program developed by Dr. Jürgen Mann is fully automated and took about three and a half hours for each seismic line. As a result of the 2D CRS stack, one obtains a simulated ZO section with high S/N ratio, an automated approximate time migration (see Mann et al., 2000, for details), a coherence section, and three kinematic wavefield attribute sections for each line.

- the tomographic velocity determination based on CRS attributes developed by Eric Duveneck took about one hour. The picking process is fully automated but needs, of course, some human interaction in order to check the result and to discard picks associated with multiple reflections.

- the calculation of the Green's function tables for migration strongly depends on the utilized method. Tables containing only traveltime information were produced in about five minutes (each line). The calculation of tables with additional dynamic information for a true-amplitude migration needs, of course, more computational effort and is, thus, more time-consuming. The

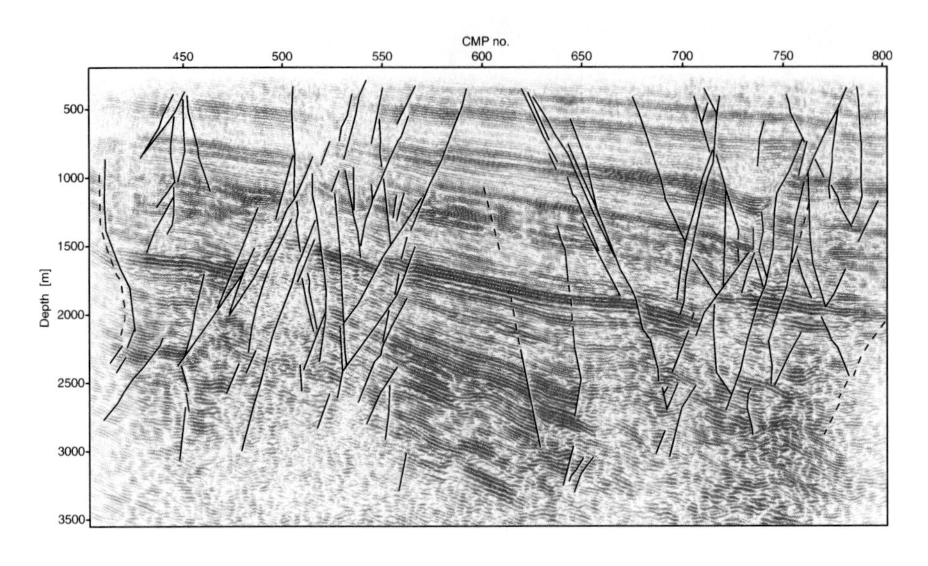

(a) Seismic line A

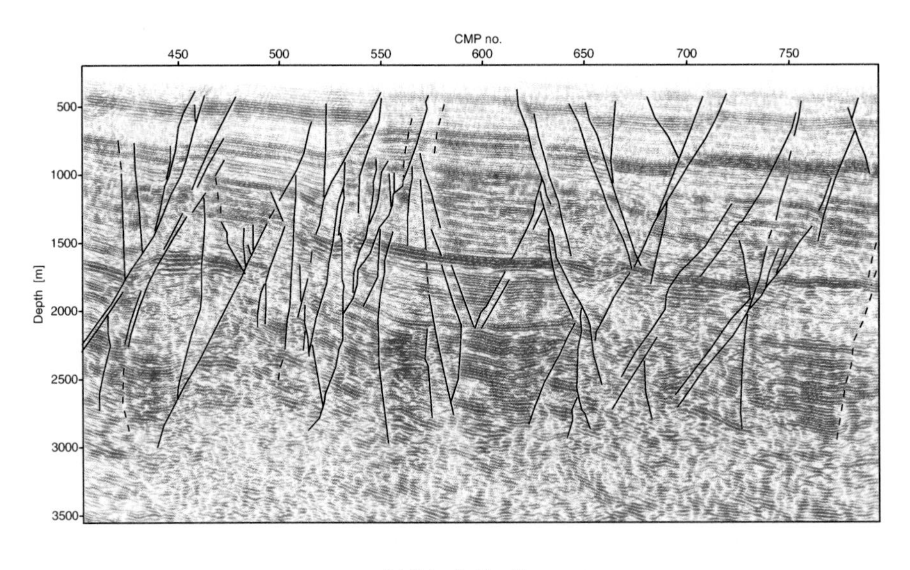

(b) Seismic line B

Figure 9.11: Preliminary structural interpretation of the prestack depth migration results. [courtesy of N. Harthill, HotRock EWK Offenbach/Pfalz GmbH]

actual runtime strongly depends on the ray-tracing technique and the spacings of the grid on which the parameters are stored, see Chapter 7 for details.

- the poststack Kirchhoff depth migration by means of the program Uni3D developed by myself took about one minute for each seismic line. This is negligible compared to the overall processing time. The prestack Kirchhoff depth migration by means of Uni3D took about 45 minutes. Offsets up to 2,000 m were considered.

In this data example, only single seismic lines were processed, i. e., a 2.5D processing scheme was applied. Although processing of 3D data is much more time-consuming, it will increasingly gain in relevance. In the hydrocarbon industry, 3D data processing is already standard. The advantages of the 2.5D imaging workflow based on the CRS stack that has been shown in this chapter and produced convincing results will be even more attractive in 3D data processing. 3D versions of all programs that have been involved in this data example are under development or already available.

Chapter 10

Summary and conclusions

In this thesis, I have studied the transformation of seismic reflection data from the time into the depth domain which is usually denoted as seismic migration or inversion. Emphasis was put on the handling of amplitudes in the migration process because the importance of preserving seismic reflection amplitudes in processing, imaging, and inversion is widely recognized. In this context, a migration process that removes geometrical spreading effects inherent in the seismic input data is referred to as true-amplitude. After true-amplitude migration, amplitudes in the migrated section are a measure of the angle-dependent reflection coefficient. As a consequence, precise AVO/AVA analyses at selected points of a target reflector can be performed. Thus, true-amplitude migration helps to reveal much more information in the data than that apparently present.

The point of departure was a brief and general introduction (Chapter 1) to reflection seismics and the description of overall properties of seismic reflection data. Afterwards, I have presented the physical and mathematical basics that are relevant for the understanding of the theories and implementations developed and explained in this thesis. The wave equation was derived from Newton's second law of mechanics in Chapter 2. Simplifications of the general elastodynamic wave equation led to the acoustic wave equation which is frequently used in seismic prospecting for hydrocarbons as an approximation to describe the wave propagation in a solid Earth. Whatever type is used, the wave equation is of fundamental importance and forms the governing equation for seismic reflection modeling as well as migration/inversion.

As a true-amplitude reflection cannot be defined without a ray theoretical description, I have shown the main ideas of (zero-order) ray theory in Chapter 3. Ray theory is a high-frequency method to approximately describe the wave propagation process in smoothly varying inhomogeneous media. The equations that form the basis of the ray method are the eikonal and transport equations. By means of these equations, traveltimes and amplitudes for seismic waves can be determined. The ray method is frequently used in the investigation of both, the forward and the inverse problem in seismics, and was used throughout this thesis.

In Chapter 4, integral solutions to the wave equation have been presented. The solution of the forward problem led to the well-known Kirchhoff integral that is frequently used for modeling purposes. The solution of the inverse problem led to the Porter-Bojarski integral and to a method called Kirchhoff migration. By simplifying the Porter-Bojarski integral for homogeneous media, the migration formula of Schneider (1978) was derived. As a generalization of Schneider's work and Hagedoorn's classical kinematic seismic reflection mapping procedures, a complete true-amplitude Kirchhoff migration approach was introduced in Chapter 5. In a geometrically appealing way, the properties of Kirchhoff migration based on the concept of Huygens surfaces have been presented. In addition, the derivation of a weight function that is applied in the diffraction stack and causes the output to be true-amplitude was explained.

In Chapter 6, I have provided an intuitive explanation of aperture effects related to (limited-aperture) Kirchhoff migration by discussing the constructive and destructive interference during the diffraction stack in simple geometrical situations. For the first time, a relationship of the terms of the stationary-phase approximation to the actually observed migration artifacts has been established. It turned out that for practical applications one has to distinguish between two principal types of migration artifacts. These are boundary effects due to a limited survey aperture and artifacts due to a limited Kirchhoff migration operator. Both types of artifacts are mathematically equivalent and can be explained by means of the boundary terms that result from the stationary-phase analysis of the migration integral. Based on the geometrical analysis, I had a closer look at a well-known way to avoid the aperture effects: tapering.

Several aspects of migration were discussed in Chapter 7, some of them are general while others are specific for Kirchhoff migration. I have shown why the traditional Kirchhoff migration method is still frequently applied and a workhorse in seismic data processing. This method needs, as any other depth migration method, a macrovelocity model with which to perform the migration process. The estimation of such a velocity model, the determination of Green's functions in the macrovelocity model, and the verification of the velocity model by analyzing CIGs after prestack migration have been addressed. If amplitudes of events are extracted from CIGs, one performs an AVO/AVA analysis. However, this information is only reliable or even possible if there exist high-quality data where amplitudes are treated in a true-amplitude sense throughout the whole processing sequence. This also implies that topographic variations of the measurement surface are properly considered in the migration process and that all aliasing effects due to the discrete nature of data acquisition and data processing are avoided as far as possible. I have shown how these problems, usually dealt with in the context of acquisition footprints and aliasing, can be solved.

The implementation of the previously mentioned aspects of Kirchhoff migration by means of a true-amplitude migration program developed in the course of this thesis was demonstrated on a synthetic data example in Chapter 8. Pre- and poststack datasets were modeled that formed the input for various migration tests. The depth migration results show that the kinematic and dynamic aspects of the Kirchhoff migration process have correctly been handled. The true-amplitude weighting functions were able to achieve the goal of removing geometrical spreading effects in the diffraction stack migration, and the handling of topographic variations of the measurement surface as well as the way to address the problem of acquisition footprints proved to be reasonable solutions. In this way, distortions and fluctuations of amplitudes in the migrated images have been minimized and, thus, amplitude analyses after migration became more reliable.

In Chapter 9, I integrated the Kirchhoff migration method into a data-driven seismic reflection imaging workflow. It was demonstrated that the Common-Reflection-Surface stack and the associated kinematic wavefield attributes can be used in seismic imaging applications which go far beyond the purposes for which the method was originally designed—the simulation of ZO sections with improved S/N ratio. The kinematic wavefield attributes have been used for the estimation of a macrovelocity model which was subsequently used in the Kirchhoff depth migration process. The workflow was applied to a real data example and led to very convincing results. Further investigations still need to be carried out to tap the full potential of this approach, especially for 3D data.

In this thesis, I have focused on the link between analytical and geometrical descriptions of Kirchhoff migration. On the basis of this relationship, I was able to close a gap between the originally graphical migration schemes of Hagedoorn (1954) and the analytical descriptions based on the work of Schneider (1978). It turned out that relating the mathematics of Kirchhoff migration to its geometrical properties is extremly helpful: on the one hand, it allows to gain an intuitive understanding of the imaging step that transforms the data from the time into the depth domain in order to provide a detailed image of the subsurface, both kinematically and dynamically. On the other hand, it helps to evaluate the results of the migration process and, thus, assists in an interpretation of the data.

Appendix A

The Method of Stationary Phase

In this Appendix, I shortly summarize the evaluation of integral (6.2) by means of the Method of Stationary Phase. For simplicity, I assume that there is only one simple point of stationary phase, denoted by ξ^*, in the integration interval (a_1, a_2). Here, "simple" means that the second derivative of q at ξ^* does not vanish, i. e., $q''(\xi^*) \neq 0$. Of course, it also must not become prohibitively small. For more than one point of stationary phase, additional integral limits are introduced, separating $I(\omega)$ into several integrals with one point ξ^* each. Then the same analysis provides the sum of contributions from these points, provided they are *isolated* from each other, i. e., each being located outside the first Fresnel zone of the others. If this is not the case, $q''(\xi^*)$ will become too small. For details about these conditions, please refer to Bleistein (1984).

As we expect the main contribution to integral (6.2) to stem from the vicinity of ξ^*, we expand $f(\xi)$ and $q(\xi)$ in Taylor series up to second order at ξ^*, where we know that $q'(\xi^*) = 0$. This approximates the integral (6.2) by a sum of three integrals,

$$
\begin{aligned}
I(\omega) \simeq\; & f(\xi^*)e^{i\omega q(\xi^*)} \underbrace{\int_{a_1}^{a_2} e^{i\alpha(\xi-\xi^*)^2} d\xi}_{I_0(\alpha)} \;+\; f'(\xi^*)e^{i\omega q(\xi^*)} \underbrace{\int_{a_1}^{a_2} (\xi-\xi^*) e^{i\alpha(\xi-\xi^*)^2} d\xi}_{I_1(\alpha)} \\
& + \frac{1}{2} f''(\xi^*)e^{i\omega q(\xi^*)} \underbrace{\int_{a_1}^{a_2} (\xi-\xi^*)^2 e^{i\alpha(\xi-\xi^*)^2} d\xi}_{I_2(\alpha)}\,, \quad \text{where} \quad \alpha = \frac{\omega}{2} \frac{d^2 q}{d\xi^2}\bigg|_{\xi^*} .
\end{aligned}
\tag{A.1}
$$

The quality of this approximation is illustrated in Figure A.1. Note that in the center part, where we expect the main contribution to integral (6.2), the coincidence is almost perfect. We remark that at non-simple stationary points, i. e., where $q''(\xi^*) = 0$, this approach will not work. In this case, the Taylor series for q and f have to be continued up to the order of the first non-vanishing derivative of q. If q is constant or zero, integral (6.2) is no longer of oscillatory character and cannot be treated by the Method of Stationary Phase.

As opposed to the integral in equation (6.2), integrals I_0, I_1, and I_2 in equation (A.1) *can* be solved analytically.

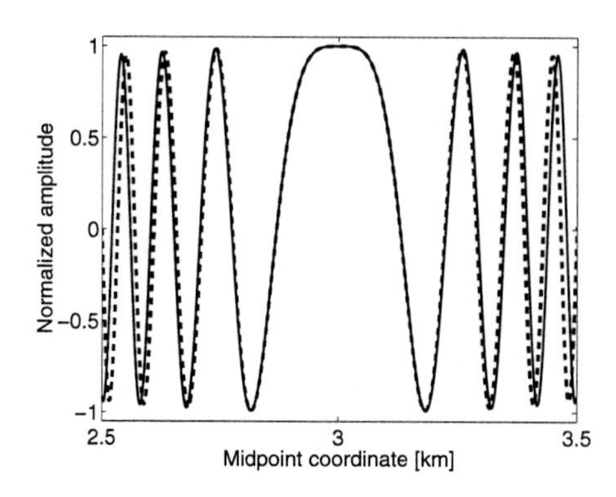

Figure A.1: Quality of the approximation of the integrand of equation (6.2) by a second-order Taylor series expansion. Shown are the real parts of the integrand function (solid line) and its approximation using second-order Taylor expansions of phase and amplitude (dashed line).

Integral I_1 is the one that is solved most easily. It yields

$$I_1(\alpha) = \frac{1}{2i\alpha} \int_{a_1}^{a_2} \frac{d}{d\xi} e^{i\alpha(\xi-\xi^*)^2} d\xi = \frac{1}{2i\alpha} \left[e^{i\alpha(\xi-\xi^*)^2} \right]_{a_1}^{a_2} , \tag{A.2}$$

where $\alpha \neq 0$ because of the condition that the stationary point must be isolated and simple, i. e., $q''(\xi^*) \neq 0$. We see that I_1 contains only contributions from the boundaries of the integration interval. Integral I_2 is immediately known once I_0 is determined since it is related to the latter as

$$I_2(\alpha) = \frac{1}{i} \int_{a_1}^{a_2} \frac{d}{d\alpha} e^{i\alpha(\xi-\xi^*)^2} d\xi = \frac{1}{i} \frac{dI_0(\alpha)}{d\alpha} . \tag{A.3}$$

The Fresnel integral I_0 requires the most extensive analysis. To study its integral limits separately, we subdivide it again into a sum of three integrals,

$$I_0(\alpha) = \int_{-\infty}^{\infty} e^{i\alpha(\xi-\xi^*)^2} d\xi - \int_{-\infty}^{a_1} e^{i\alpha(\xi-\xi^*)^2} d\xi - \int_{a_2}^{\infty} e^{i\alpha(\xi-\xi^*)^2} d\xi . \tag{A.4}$$

In case the stationary point is at (or very close to) one of the integral boundaries a_1 or a_2, the corresponding one of the above boundary integrals is eliminated and the first integral is carried out from ξ^* to infinity.

As the phase function is monotonic in the two boundary integrals, a transformation of variables $(\xi - \xi^*)^2 = u$ and subsequent repeated partial integration yields a power series in $1/\alpha$ (or $1/\omega$), the leading terms of which are

$$\int_{-\infty}^{a_1} e^{i\alpha(\xi-\xi^*)^2} d\xi \simeq \frac{1}{2i\alpha} \frac{1}{(a_1 - \xi^*)} e^{i\alpha(a_1-\xi^*)^2} , \tag{A.5}$$

and

$$\int_{a_2}^{\infty} e^{i\alpha(\xi-\xi^*)^2} d\xi \simeq -\frac{1}{2i\alpha} \frac{1}{(a_2 - \xi^*)} e^{i\alpha(a_2-\xi^*)^2} . \tag{A.6}$$

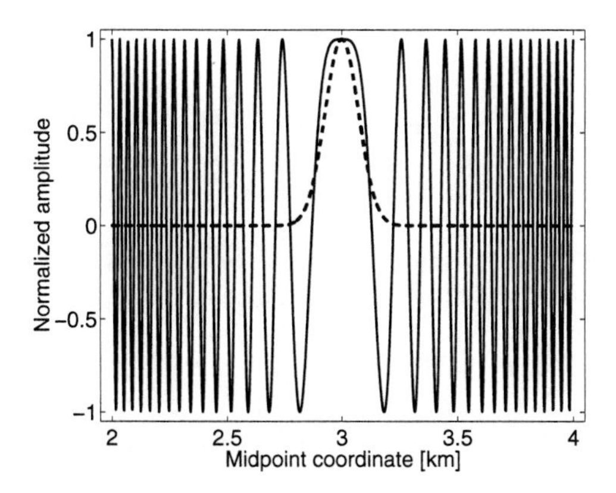

Figure A.2: Real part of the oscillating (solid line) and bell-shaped (dashed line) integrand functions of the integrals in equation (A.7).

For non-isolated points of stationary phase, the distance between ξ^* and at least one of the integral boundaries is too small, such that the corresponding approximation (A.5) or (A.6) is not valid.

The integrals in equations (A.5) and (A.6) are the boundary contributions to I_0. The remaining integral describes the contribution from the point of stationary phase. A detailed analysis in the complex plane (see, e. g., Stamnes, 1986) shows that

$$\int_{-\infty}^{\infty} e^{i\alpha(\xi-\xi^*)^2} d\xi \;=\; \frac{1+i\,\mathrm{sgn}\,\alpha}{\sqrt{2}} \int_{-\infty}^{\infty} e^{-|\alpha|\xi^2} d\xi = \sqrt{\frac{\pi}{|\alpha|}}\, e^{i\frac{\pi}{4}\,\mathrm{sgn}\,\alpha} \;=\; \sqrt{\frac{\pi}{-i\alpha}}\,, \tag{A.7}$$

with the definition of the complex square root

$$\sqrt{z} = \sqrt{|z|}\exp\{i\arg(z)/2\}\,, \qquad -\pi < \arg(z) \le \pi\,. \tag{A.8}$$

By symmetry, the left-hand-side integral in equation (A.7) yields exactly half this contribution if its lower limit is ξ^*.

Figure A.2 visualizes the above identity (A.7). The real part of the oscillating function $e^{i\alpha(\xi-\xi^*)^2}$, and the bell-shaped function $e^{-|\alpha|\xi^2}$ are the solid and dashed curves, respectively. Note that equation (A.7) states that integrations from minus infinity to infinity over the two curves in Figure A.2 yield identical results, except for a factor $1/\sqrt{2}$, or, considering also the imaginary parts, $e^{i\frac{\pi}{4}\,\mathrm{sgn}\,\alpha}$.

Combining equations (A.5) to (A.7), we obtain for I_0 up to the first order in α^{-1}

$$I_0(\alpha) \;\simeq\; \sqrt{\frac{\pi}{-i\alpha}} + \frac{1}{2i\alpha}\left(\frac{1}{a_2-\xi^*}e^{i\alpha(a_2-\xi^*)^2} - \frac{1}{a_1-\xi^*}e^{i\alpha(a_1-\xi^*)^2}\right). \tag{A.9}$$

By equation (A.3), this yields for I_2 up to the first order in α^{-1}

$$I_2 \simeq \frac{1}{2i\alpha}\left((a_2-\xi^*)e^{i\alpha(a_2-\xi^*)^2} - (a_1-\xi^*)e^{i\alpha(a_1-\xi^*)^2}\right) \tag{A.10}$$

In other words, like I_1, I_2 also describes only contributions from the boundaries of the integration interval to that order. Collecting the terms of equation (A.1) and recognizing the Taylor expansions of

$\frac{dq}{d\xi}$ and $f(\xi)$, we finally find

$$I(\omega) \simeq f(\xi^*)e^{i\omega q(\xi^*)}\sqrt{\frac{2\pi}{-i\omega q''(\xi^*)}} + \frac{1}{i\omega}\left[\frac{f(a_2)}{q'(a_2)}e^{i\omega q(a_2)} - \frac{f(a_1)}{q'(a_1)}e^{i\omega q(a_1)}\right], \tag{A.11}$$

where the prime denotes the derivative with respect to ξ. Under the assumption of a single, isolated point of stationary phase, the above analysis of the migration integral (6.1) by means of the Method of Stationary Phase has shown that its first two terms in equation (A.11) are of the order $1/\sqrt{\omega}$ and $1/\omega$, respectively. If the stationary point coincides with one of the boundaries, i. e., if $q'(a_1) = 0$ or $q'(a_2) = 0$, equation (A.11) has to be slightly modified. The corresponding boundary contribution at a_1 or a_2 is eliminated and the leading term is divided by 2, as already indicated in the context of equations (A.4) and (A.7).

If there are two simple, isolated stationary points of the phase function $q(\xi)$ in the integration interval $[a_1, a_2]$, the interval is divided at some intermediate point c into $[a_1, c]$ and $[c, a_2]$. Then

$$I(\omega) = \int_{a_1}^{a_2} f(\xi)e^{i\omega q(\xi)}d\xi = \int_{a_1}^{c} f(\xi)e^{i\omega q(\xi)}d\xi + \int_{c}^{a_2} f(\xi)e^{i\omega q(\xi)}d\xi\,. \tag{A.12}$$

The independent analysis of each of the two separate integrals can then be carried out as before, yielding as the final result

$$\begin{aligned} I(\omega) \simeq\; & f(\xi_1^*)e^{i\omega q(\xi_1^*)}\sqrt{\frac{2\pi}{-i\omega q''(\xi_1^*)}} + \frac{1}{i\omega}\left[\frac{f(c)}{q'(c)}e^{i\omega q(c)} - \frac{f(a_1)}{q'(a_1)}e^{i\omega q(a_1)}\right] \\ & + f(\xi_2^*)e^{i\omega q(\xi_2^*)}\sqrt{\frac{2\pi}{-i\omega q''(\xi_2^*)}} + \frac{1}{i\omega}\left[\frac{f(a_2)}{q'(a_2)}e^{i\omega q(a_2)} - \frac{f(c)}{q'(c)}e^{i\omega q(c)}\right]. \end{aligned} \tag{A.13}$$

We see that the artificially introduced integral boundary at c does not contribute to the final value of $I(\omega)$ as the corresponding boundary terms cancel each other. Correspondingly, for N simple, isolated stationary points within $[a_1, a_2]$, evaluation of integral (6.2) yields the sum of their contributions,

$$I(\omega) \simeq \sum_{n=1}^{N} f(\xi_n^*)e^{i\omega q(\xi_n^*)}\sqrt{\frac{2\pi}{-i\omega q''(\xi_n^*)}} + \frac{1}{i\omega}\left[\frac{f(a_2)}{q'(a_2)}e^{i\omega q(a_2)} - \frac{f(a_1)}{q'(a_1)}e^{i\omega q(a_1)}\right]. \tag{A.14}$$

Appendix B

Projected Fresnel zone

In this Appendix, I derive expression (6.10) for the projected Fresnel zone in the zero-offset configuration, assuming a planar reflector with dip θ and a constant background velocity v (see Figure B.1). The projected Fresnel zone is defined as the projection of the true Fresnel zone in depth along neighboring reflection rays onto the Earth's surface (Hubral et al., 1993). In other words, the projected Fresnel zone ends where the rays reflected at the boundaries of the true Fresnel zone reach the Earth's surface.

I start from the definition of the Fresnel zone, equation (6.3). At ξ^*, the reflection traveltime, τ_R, of the normal ray reflected at M (see Figure B.1), is

$$\tau_R = \frac{2}{v} \sqrt{r^2 + z^2} \,, \tag{B.1}$$

where r is given by equation (6.9). The diffraction traveltime of a neighboring reflector point $\overline{M}(n)$, also measured at ξ^*, is

$$\tau_D = \frac{2}{v} \sqrt{r^2 + z^2 + \ell(n)^2} \,, \tag{B.2}$$

where $\ell(n)$ is the distance between M and $\overline{M}(n)$. Substituting these two expressions for τ_R and τ_D in equation (6.3) and solving for $\ell(n)$, one finds

$$\ell(n) = \sqrt{\left(\frac{vnT}{4}\right)^2 + \frac{vnT}{2} \sqrt{r^2 + z^2}} = \sqrt{\left(\frac{vnT}{4}\right)^2 + \frac{vnTz}{2\cos\theta}} \,. \tag{B.3}$$

This is the size of the true Fresnel zone at the reflector in depth. To obtain the size of the projected Fresnel zone, I still have to project this distance onto the Earth's surface along neighboring normal rays (dashed rays in Figure B.1). Since these rays are parallel, the projection provides an additional division by $\cos\theta$, thus yielding formula (6.10).

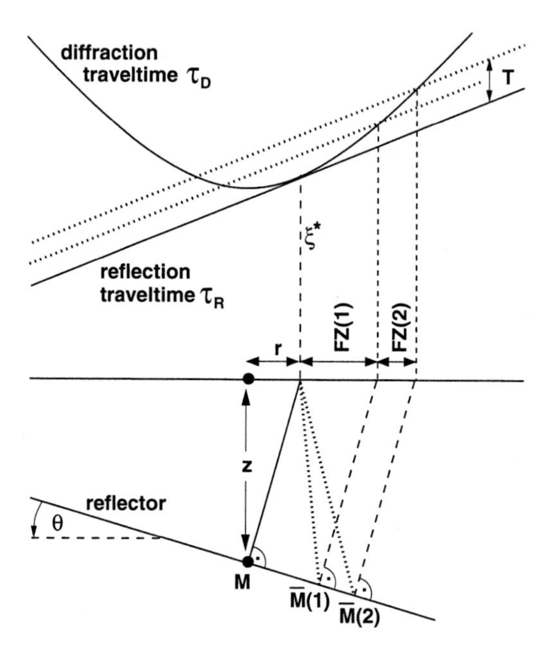

Figure B.1: Construction of the projected Fresnel zone. The parameter ξ^* denotes the stationary point, i. e., the point where the reflection traveltime curve τ_R and the diffraction traveltime τ_D have the same slope. The distance between the vertical projection of M onto the surface and the stationary point is denoted by r. The projection of the (first) Fresnel zone $M - \overline{M}(1)$ in the depth domain onto the Earth's surface yields the (first) projected Fresnel zone $FZ(1)$. The dip of the reflector is given by θ, T denotes the period of the monofrequency wave. It may be replaced by some estimate of the wavelet length τ_w. The construction of the second projected Fresnel zone proceeds similarly.

List of Figures

Bibliography

Abma, R., Sun, J., and Bernitsas (1999). Antialiasing methods in Kirchhoff migration. *Geophysics*, 64:1783–1792.

Aki, R. and Richards, P. (1980). *Quantitative Seismology: Theory and methods*, volume 1. Freeman and Company, New York.

Al-Yahya, K. (1989). Velocity analysis by iterative profile migration. *Geophysics*, 54:718–729.

Audebert, F., Nichols, D., Rekdal, T., Biondi, B., Lumley, D., and Urdaneta, H. (1997). Imaging complex geologic structure with single-arrival Kirchhoff prestack depth migration. *Geophysics*, 62:1533–1543.

Baina, R., Nguyen, S., Noble, M., and Lambaré, G. (2003). Optimal anti-aliasing for ray-based Kirchhoff depth migration. In *Expanded Abstracts*. 73rd Ann. Internat. Mtg., Soc. Expl. Geophys. Session MIG P2.2.

Bergler, S., Hubral, P., Marchetti, P., Cristini, A., and Cardone, G. (2002). 3D common-reflection-surface stack and kinematic wavefield attributes. *The Leading Edge*, 21:1010–1015.

Berkhout, A. (1987). *Applied Seismic Wave Theory*. Advances in Exploration Geophysics. Elsevier, Amsterdam.

Bevc, D. (1997). Flooding the topography: Wave-equation datuming of land data with rugged acquisition topography. *Geophysics*, 62:1558–1569.

Beydoun, W., Hanitzsch, C., and Jin, S. (1993). Why migrate before AVO? In *Extended Abstracts*. 55th Conference & Exhibition, Europ. Assoc. Geosci. Eng. Session B044.

Beylkin, G. (1985a). Imaging of discontinuities in the inverse scattering problem by inversion of a generalized Radon transform. *J. Math. Phys.*, 26:99–108.

Beylkin, G. (1985b). Reconstructing discontinuities in multidimensional inverse scattering problems. *Applied Optics*, 24:4086–4088.

Billette, F., Etgen, J., and Rietveld, W. (2000). Prestack depth migration using fast 3D tomography. In *Extended Abstracts*. 62nd Conference & Exhibition, Europ. Assoc. Geosci. Eng. Session L0037.

Biloti, R., Santos, L., and Tygel, M. (2002). Multiparametric traveltime inversion. *Stud. geophys. geod.*, 46:177–192.

Biondi, B. (2001). Kirchhoff imaging beyond aliasing. *Geophysics*, 66(2):654–666.

Biondi, B. (2003). 3-D Seismic Imaging. Lecture notes, Stanford University. http://sepwww.stanford.edu/sep/biondo/.

Bleistein, N. (1984). *Mathematical Methods for Wave Phenomena.* Academic Press Inc., New York.

Bleistein, N. (1987). On the imaging of reflectors in the earth. *Geophysics*, 52:931–942.

Bleistein, N. (1999). Hagedoorn told us how to do Kirchhoff migration and inversion. *The Leading Edge*, 18:918–927.

Bleistein, N., Cohen, J., and Stockwell Jr., J. (2001). *Mathematics of Multidimensional Seismic Imaging, Migration, and Inversion.* Springer Verlag, New York.

Bleistein, N. and Handelsman, R. (1986). *Asymptotic expansions of integrals.* Dover Publications Inc., New York.

Bortfeld, R. (1989). Geometrical ray theory: Rays and traveltimes in seismic systems. *Geophysics*, 54:342–349.

Bronstein, I. and Semendjajew, K. (1997). *Handbook of Mathematics.* Springer Verlag, New York.

Canning, A. and Gardner, G. (1998). Reducing 3-D acquisition footprint for 3-D DMO and 3-D prestack migration. *Geophysics*, 63:1177–1183.

Červený, V. (2001). *Seismic Ray Theory.* Cambridge Univ. Press, Cambridge.

Červený, V. and Hron, F. (1980). The ray series method and dynamic ray-tracing system for three-dimensional inhomogeneous media. *Bull. Seis. Soc. Am.*, 70:47–77.

Červený, V., Popov, M., and Pšenčík, I. (1982). Computation of Wave Fields in Inhomogeneous Media: Gaussian Beam Approach. *Geophys. J. R. Astr. Soc.*, 70:109–128.

Červený, V. and Soares, J. (1992). Fresnel volume ray tracing. *Geophysics*, 57:902–915.

Chauris, H., Noble, M., Lambaré, G., and Podvin, P. (2002). Migration velocity analysis from locally coherent events in 2-D laterally heterogeneous media, Part I: theoretical aspects. *Geophysics*, 67:1202–1212.

Claerbout, J. (1971). Toward a unified theory of reflector mapping. *Geophysics*, 36:467–481.

Claerbout, J. (1992a). Anti aliasing. Stanford exploration project report 73, Stanford University.

Claerbout, J. (1992b). *Earth sounding analysis: Processing versus inversion.* Blackwell Scientific Publications, Boston.

Crawley, S. (1996). Antialiasing 3D prestack Kirchhoff datuming. Stanford exploration project report 93, Stanford University.

Cristini, A., Cardone, G., and Marchetti, P. (2002). 3D zero-offset Common Reflection Surface Stack for land data – real data example. In *Extended Abstracts.* 64th Conference & Exhibition, Europ. Assoc. Geosci. Eng.

Delaunay, B. (1934). Sur la sphère vide. *Otdelenie Matematicheskii i Estestvennyka Nauk*, 7:793–800.

Docherty, P. (1991). A brief comparison of some Kirchhoff integral formulas for migration and inversion. *Geophysics*, 56:1164–1169.

Duveneck, E. (2004). Velocity model estimation with data-derived wavefront attributes. *Geophysics*, 69. In print.

Duveneck, E. and Hubral, P. (2002). Tomographic velocity model inversion using kinematic wavefield attributes. In *Expanded Abstracts*. 72nd Ann. Internat. Mtg., Soc. Expl. Geophys. Session IT 2.3.

Fagin, S. (1999). Model-Based Depth Imaging. In Young, R., editor, *Course Note Series, No. 10*. Soc. Expl. Geophys., Tulsa.

Farmer, P., Gray, S., Hodgkiss, G., Pieprzak, A., Ratcliff, D., and Whitcombe, D. (1993). Structural Imaging: Toward a Sharper Subsurface View. *Oilfield Review*, 5:28–41.

Felsen, L. and Marcuvitz, N. (1973). *Radiation and Scattering of Waves*. Prentice-Hall Inc., Englewood Cliffs.

Fink, M., Kuperman, W., Montagner, J.-P., and Tourin, A. (2002). *Imaging of Complex Media with Acoustic and Seismic Waves*. Topics in Applied Physics Vol. 82. Springer Verlag, New York.

Gesbert, S. (2002). From acquisition footprints to true amplitude. *Geophysics*, 62:830–839.

Goertz, A. (2002). *True-amplitude multicomponent migration of elastic wavefields*. Logos Verlag, Berlin.

Gold, N., Shapiro, S., Bojinski, S., and Müller, T. (2000). An approach to upscaling for seismic waves in statistically isotropic heterogeneous elastic media. *Geophysics*, 65:1837–1850.

Gray, S. (1992). Frequency-selective design of the Kirchhoff migration operator. *Geophys. Prosp.*, 40:565–571.

Gray, S. (1997). True-amplitude seismic migration: A comparison of three approaches. *Geophysics*, 62:929–936.

Gray, S. (1998). Speed and accuracy of seismic migration methods. Mathematical Geophysics Summer School (Proceedings), Stanford University. http://sepwww.stanford.edu/etc/sam_gray/.

Gray, S., Etgen, J., Dellinger, J., and Whitmore, D. (2001). Seismic migration problems and solutions. *Geophysics*, 66:1622–1640.

Gray, S., MacLean, G., and Marfurt, K. (1999). Crooked line, rough topography: Advancing towards the correct seismic image. *Geophys. Prosp.*, 47:721–733.

Gray, S. and Marfurt, K. (1995). Migration from Topography: Improving the Near-Surface Image. *Can. J. Expl. Geophys.*, 31:18–24.

Grubb, H. and Walden, A. (1995). Smoothing seismically derived velocities. *Geophys. Prosp.*, 43:1061–1082.

Hagedoorn, J. (1954). A process of seismic reflection interpretation. *Geophys. Prosp.*, 2:85–127.

Hanitzsch, C. (1997). Comparison of weights in prestack amplitude-preserving Kirchhoff depth migration. *Geophysics*, 62:1812–1816.

Hanyga, A. (1984). Seismic Wave Propagation in the Earth. In Hanyga, A., editor, *Physics and Evolution of the Earth's Interior 2*. Elsevier, Amsterdam.

Hatchell, P. (2000). Fault whispers: Transmission distortions on prestack seismic reflection data. *Geophysics*, 65:377–389.

Herman, G., Langenberg, K., and Sabatier, P. (1986). *Basic Methods of Tomography and Inverse Problems*. Malvern Physics Series. Adam Hilger, Bristol.

Hill, N. (1990). Gaussian beam migration. *Geophysics*, 55:1416–1428.

Hill, N. (2001). Prestack Gaussian-beam depth migration. *Geophysics*, 66:1240–1250.

Höcht, G. (2002). *Traveltime approximations for 2D and 3D media and kinematic wavefield attributes*. PhD thesis, University of Karlsruhe.

Hubral, P., editor (1999). *Macro-model independent seismic reflection imaging*, volume 42(3,4). J. Appl. Geophys.

Hubral, P., Schleicher, J., and Tygel, M. (1996). A unified approach to 3-D seismic reflection imaging, Part I: Basic concepts. *Geophysics*, 61:742–758.

Hubral, P., Schleicher, J., Tygel, M., and Hanitzsch, C. (1993). Determination of Fresnel zones from traveltime measurements. *Geophysics*, 58:703–712.

Hubral, P., Schleicher, S., and Tygel, M. (1992). Three-dimensional Paraxial Ray Properties. Part I: Basic Relations. *J. Seis. Expl.*, 1:265–279.

Hubral, P., Tygel, M., and Schleicher, J. (1995). Geometrical-spreading and ray-caustic decomposition of elementary seismic waves. *Geophysics*, 60:1195–1202.

Jäger, C., Hertweck, T., and Spinner, M. (2003). True-amplitude Kirchhoff migration from topography. In *Expanded Abstracts*. 73rd Ann. Internat. Mtg., Soc. Expl. Geophys. Session MIG 2.1.

Jäger, R., Mann, J., Höcht, G., and Hubral, P. (2001). Common-Reflection-Surface stack: Image and attributes. *Geophysics*, 66:97–109.

James, G. (1976). *Geometrical Theory of Diffraction for Electromagnetic Waves*. Peter Peregrinus Ltd., Stevenage.

Jaramillo, H., Schleicher, J., and Tygel, M. (1998). Discussion and Errata to: A unified approach to 3-D seismic reflection imaging, Part II: Theory. *Geophysics*, 63:670–673.

Jin, S. and Madariaga, R. (1993). Background velocity inversion with a genetic algorithm. *Geophys. Res. Lett.*, 20:93–96.

Jin, S. and Madariaga, R. (1994). Nonlinear velocity inversion by a two-step Monte Carlo method. *Geophysics*, 59:577–590.

Jones, I. and Fruehn, J. (2003). Factors affecting frequency content in PreSDM imaging. In *Extended Abstracts*. 65th Conference & Exhibition, Europ. Assoc. Geosci. Eng. Session E-07.

Jousset, P., Thierry, P., and Lambaré, G. (1999). Reduction of 3-D acquisition footprints in 3-D migration/inversion. In *Expanded Abstracts*. 70th Ann. Internat. Mtg., Soc. Expl. Geophys. Session SPRO 11.1.

Jousset, P., Thierry, P., and Lambaré, G. (2000). Improvement of 3D Migration/inversion by reducing acquisition footprints: application to real data. In *Expanded Abstracts*. 70th Ann. Internat. Mtg., Soc. Expl. Geophys. Session MIG 5.5.

Kravtsov, Y. and Orlov, Y. (1990). *Geometrical Optics of Inhomogeneous Media*. Springer Verlag, New York.

Lay, T. and Wallace, T. (1995). *Modern Global Seismology*. Academic Press, San Diego.

Liner, L. (1999). Concepts of normal and dip moveout. *Geophysics*, 64:1637–1647.

Lumley, D., Claerbout, J., and Bevc, D. (1994). Anti-aliased Kirchhoff 3-D migration. In *Expanded Abstracts*, pages 1282–1285. 64th Ann. Internat. Mtg., Soc. Expl. Geophys.

Luneburg, R. (1966). *Mathematical Theory of Optics*. Univ. of California Press, Berkeley.

Majer, P. (2000). Inversion of seismic parameters: Determination of the 2-D iso-velocity layer model. Master's thesis, University of Karlsruhe.

Mann, J. (2002). *Extensions and Applications of the Common-Reflection-Surface Stack Method*. Logos Verlag, Berlin.

Mann, J. and Höcht, G. (2003). Pulse stretch effects in the context of data-driven imaging methods. In *Extended Abstracts*. 65th Conference & Exhibition, Europ. Assoc. Geosci. Eng. Session P007.

Mann, J., Hubral, P., Traub, B., Gerst, A., and Meyer, H. (2000). Macro-Model Independent Approximative Prestack Time Migration. In *Extended Abstracts*. 62nd Conference & Exhibition, Europ. Assoc. Geosci. Eng. Session B-52.

Martins, J., Schleicher, J., Tygel, M., and Santos, L. (1997). 2.5-D True-amplitude Migration and Demigration. *J. Seis. Expl.*, 6:159–180.

Müller, T. (1998). Common Reflection Surface Stack versus NMO/Stack and NMO/DMO/Stack. In *Extended Abstracts*. 60th Conference & Exhibition, Europ. Assoc. Geosci. Eng. Session 1-20.

Müller, T. (1999). *The Common Reflection Surface Stack Method – Seismic imaging without explicit knowledge of the velocity model*. Der Andere Verlag, Bad Iburg.

Newman, P. (1973). Divergence effects in a layered earth. *Geophysics*, 38:481–488.

Newman, P. (1975). Amplitude and Phase Properties of a Digital Migration Process. In *Extended Abstracts*. Europ. Assoc. Expl. Geoph. (Republished in: First Break, 8:397-403, 1990).

Peres, O., Kosloff, D., Koren, Z., and Tygel, M. (2001). A practical approach to true-amplitude migration. *J. Seis. Expl.*, 10:183–204.

Popov, M. (2002). *Ray Theory and Gaussian Beam Method for Geophysicists*. Editora da Universidade Federal da Bahia, Salvador.

Popov, M. and Pšenčík, I. (1978). Computation of ray amplitudes in inhomogeneous media with curved interfaces. *Studia geoph. et geod.*, 22:248–258.

Robein, E. (2003). *Velocities, Time-imaging and Depth-imaging in Reflection Seismics. Principles and Methods*. EAGE Publications, Houten.

Rohr, K., Scheidhauer, M., and Trehu, A. (2000). Transpression between two warm mafic plates: the Queen Charlotte Fault revisited. *J. Geophys. Res.*, 105:8147–8172.

Scales, J. (1997). *Theory of Seismic Imaging*. Samizdat Press, Golden.

Scheidhauer, M., Trehu, A., and Rohr, K. (1999). Multi-channel seismic reflection survey over the northern Queen Charlotte Fault, offshore British Columbia. Open File Report 3779, Geological Survey of Canada.

Schleicher, J. (1993). *Bestimmung von Reflexionskoeffizienten aus Reflexionsseismogrammen*. PhD thesis, University of Karlsruhe.

Schleicher, J., Hubral, P., Tygel, M., and Jaya, M. (1997). Minimum apertures and Fresnel zones in migration and demigration. *Geophysics*, 62:183–194.

Schleicher, J., Tygel, M., and Hubral, P. (1993). 3-D true-amplitude finite-offset migration. *Geophysics*, 58:1112–1126.

Schneider, W. (1978). Integral Formulation for Migration in Two and Three Dimensions. *Geophysics*, 43:49–76.

Sheriff, R. (1975). Factors affecting seismic amplitudes. *Geophys. Prosp.*, 23:125–138.

Shuey, R. (1985). A simplification of the Zoeppritz equations. *Geophysics*, 50:607–614.

Smith, M. (1993). *Theoretical Seismology*. Samizdat Press, Vermont.

Spinner, M. (2003). True-amplitude Kirchhoff migration from topography – theory and application. Master's thesis, Karlsruhe University.

Spitz, S. (1991). Seismic trace interpolation in the f-x domain. *Geophysics*, 56:785–794.

Stamnes, J. (1986). *Waves in Focal Regions*. Adam Hilger, Bristol and Boston.

Sullivan, M. and Cohen, J. (1987). Prestack Kirchhoff inversion of common-offset data. *Geophysics*, 62:745–754.

Sun, J. (1998). On the limited aperture migration in two dimensions. *Geophysics*, 63:984–994.

Sun, J. (1999). On the aperture effect in 3-D Kirchhoff-type migration. *Geophys. Prosp.*, 47:1045–1076.

Sun, J. (2000). Limited aperture migration. *Geophysics*, 65:584–595.

Sun, J. and Gajewski, D. (1997). True-amplitude common-shot migration revisited. *Geophysics*, 62:1250–1259.

Sun, J. and Gajewski, D. (1998). On the computation of the true-amplitude weighting functions. *Geophysics*, 63:1648–1651.

Sun, S. and Bancroft, J. (2001). How much does the migration aperture actually contribute to the migration result? In *Expanded Abstracts*, pages 973–976. 71th Ann. Internat. Mtg., Soc. Expl. Geophys.

Taner, M., Koehler, F., and Sheriff, R. (1979). Complex seismic trace analysis. *Geophysics*, 44:1041–1063.

Trappe, H., Gierse, G., and Pruessmann, J. (2001). Case studies show potential of Common Reflection Surface stack - structural resolution in the time domain beyond the conventional NMO/DMO stack. *First Break*, 19:625–633.

Trappe, H., Gierse, G., Pruessmann, J., Laggiard, E., Boennemann, C., and Meyer, H. (2003). CRS imaging of 3-D seismic data from the active continental margin offshore Costa Rica. In *Extended Abstracts*. 65th Conference & Exhibition, Europ. Assoc. Geosci. Eng.

Tygel, M., Schleicher, J., and Hubral, P. (1994a). Kirchhoff-Helmholtz theory in modelling and migration. *J. Seis. Expl.*, 3:203–214.

Tygel, M., Schleicher, J., and Hubral, P. (1994b). Pulse distortion in depth migration. *Geophysics*, 59:1561–1569.

Tygel, M., Schleicher, J., and Hubral, P. (1995). Dualities between Reflectors and Reflection-Time Surfaces. *J. Seis. Expl.*, 4:123–150.

Tygel, M., Schleicher, J., and Hubral, P. (1996). A unified approach to 3-D seismic reflection imaging, Part II: Theory. *Geophysics*, 61:759–775.

Vanelle, C. (2002). *Traveltime-based True-amplitude Migration*. PhD thesis, University of Hamburg.

Vanelle, C. and Gajewski, D. (2002a). Second-order interpolation of traveltimes. *Geophys. Prosp.*, 50:73–83.

Vanelle, C. and Gajewski, D. (2002b). True-amplitude migration weights from traveltimes. *Pure Appl. Geophys.*, 159:1583–1599.

Vermeer, G. (1990). *Seismic wavefield sampling*. Soc. Expl. Geophys., Tulsa.

Vidale, J. (1988). Finite-Difference Traveltime Calculations. *Bull. Seis. Soc. Am.*, 78:2062–2076.

Vieth, K.-U. (2001). *Kinematic wavefield attributes in seismic imaging*. PhD thesis, University of Karlsruhe.

Vinje, V., Iversen, E., Åstebøl, K., and Gjøystdal, H. (1996a). Estimation of multivalued arrivals in 3D models using wavefront construction – Part I. *Geophys. Prosp.*, 44:819–842.

Vinje, V., Iversen, E., Åstebøl, K., and Gjøystdal, H. (1996b). Estimation of multivalued arrivals in 3D models using wavefront construction – Part II: Tracing and interpolation. *Geophys. Prosp.*, 44:843–858.

Vinje, V., Iversen, E., and Gjøystdal, H. (1993). Traveltime and amplitude estimation using wavefront construction. *Geophysics*, 58:1157–1166.

Voronoj, G. (1908). Nouvelles Applications des Paramètres Continus à la Théorie des Fores. Deuxième Mémoire. Recherches sur les Parallélloèdres Primitifs. *J. Pure Appl. Math.*, 134:198–287.

Wapenaar, C. (1992). The Infinite Aperture Paradox. *J. Seis. Expl.*, 1:325–336.

Williamson, P., Berthet, P., Mispel, J., and Sexton, P. (1999). Anisotropic velocity model construction and migration: An example from West Africa. In *Expanded Abstracts*, pages 1592–1595. 69th Ann. Internat. Mtg., Soc. Expl. Geophys.

Xu, S., Chauris, H., Lambaré, G., and Noble, M. (2001). Common-angle migration: A strategy for imaging complex media. *Geophysics*, 66:1877–1894.

Yilmaz, Ö. (2001). *Seismic Data Analysis*. Soc. Expl. Geophys., Tulsa.

Zhang, Y., Gray, S., Sun, J., and Notfors, C. (2001). Theory of migration anti-aliasing. In *Expanded Abstracts*. 71st Ann. Internat. Mtg., Soc. Expl. Geophys. Session MIG 4.7.

Ziolkowski, R. and Deschamps, G. (1980). The Maslov method and the asymptotic Fourier transform: Caustic analysis. Electromagnetic Laboratory Scientific report 80-9, University of Illinois.

Zoeppritz, K. (1919). Erdbebenwellen VII: Über Reflexion und Durchgang seismischer Wellen durch Unstetigkeitsflächen. *Nachrichten der Königlichen Gesellschaft der Wissenschaften zu Göttingen*, pages 66–84.

Used hard- and software

This thesis was written on a computer with the free operating system Linux (SuSE Linux 8.2) using the word processing package TEX, the macro package LATEX 2ε, and several extensions. The bibliography was generated with BIBTEX.

For analytical or numerical calculations (and partly for visualization), the mathematical programs Matlab (The MathWorks) and Maple (Maplesoft) were used. Schematical figures were mainly constructed with Xfig.

The freely available seismic processing packages SEPlib (Stanford Exploration Project, Stanford University) and Seismic Un*x (Center for Wave Phenomena, Colorado School of Mines) were used to preprocess and visualize seismic data.

The Kirchhoff true-amplitude migration program Uni3D that was developed in the course of this thesis is written in C++. It was used to create all of the depth-migrated images presented in this thesis. The SGI and the GNU C++ compilers as well as the GNU debugger gdb with the frontend ddd were used in the development process. The program is implemented for the i386 architecture on PCs running Linux, for the RISC architecture on HP workstations running HP-UX, and for the MIPS architecture on an SGI Origin 3200 running IRIX.

Norsar 3D Ray Modelling was used as ray-tracing tool to create the synthetic data examples shown in this thesis. In addition, it served as an efficient tool to build some of the Green's function tables used in the synthetic and real data migration examples. For fast traveltime calculations, the SEPlib program FMeikonal was used.

In addition, various small public domain programs and shell scripts were used to edit, process, or visualize data.

Danksagung / Acknowledgement

Zur Entstehung der vorliegenden Arbeit haben zahlreiche Personen und Institutionen in unterschiedlicher Weise beigetragen. Es ist nahezu unmöglich, alle zu erwähnen, die sich auf ihre ganz individuelle Art und Weise eingebracht und mir geholfen haben. Die folgende Aufzählung ist daher keinesfalls vollständig. Ich möchte mich ausdrücklich bei all denen entschuldigen, die im Folgenden zu Unrecht nicht explizit genannt werden.

Prof. Peter Hubral danke ich für die Betreuung meiner Arbeit und die Übernahme des Referats. Er hat mich stets in jeder Hinsicht unterstützt und es mir ermöglicht, meine Arbeiten auf zahlreichen internationalen Tagungen zu präsentieren. Er ließ mir während meiner gesamten Doktorandenzeit viel Gestaltungsfreiraum und hatte stets ein offenes Ohr für meine Ideen. Er gab mir zudem die Gelegenheit, eine kleine Arbeitsgruppe zum Thema Migration am Geophysikalischen Institut der Universität Karlsruhe mit aufzubauen. Von der damit verbundenen Verantwortung sowie von seiner Erfahrung und Förderung konnte ich in vielerlei Hinsicht sehr profitieren.

Prof. Friedemann Wenzel danke ich für die Übernahme des Korreferats. Er hat sich einst selbst mit den Theorien um die amplitudenbewahrende Migration beschäftigt und damit einen Baustein für die Technologie geliefert, die uns heute zur Verfügung steht.

Prof. Jörg Schleicher gilt mein Dank für die gute Zusammenarbeit und seine Bereitschaft, mir alle Fragen zur amplitudenbewahrenden Migration stets geduldig zu beantworten. Durch unsere Diskussionen und von seinen Artikeln habe ich viel gelernt.

I am really thankful to **Prof. Mikhail Popov**. He showed me the secrets of ray theory and gave me many (mathematical) insights into the theory of inversion and migration. His constructive criticism and the fruitful discussions with him have helped a lot to improve this thesis, especially concerning Chapter 4. He also proofread this chapter.

Many thanks to **Prof. Norman Bleistein**. He is a specialist in true-amplitude migration and the Method of Stationary Phase and, thus, I could really profit from his stays in Karlsruhe and from several discussions. Furthermore, he reviewed Chapter 6 and gave some valuable suggestions for improvement.

Diese Arbeit wäre nicht möglich gewesen ohne die unmittelbare Unterstützung durch die Mitglieder der Arbeitsgruppe „True-amplitude Migration“ am Geophysikalischen Institut. Mit **Christoph Jäger** teilte ich nicht nur das thematische Interesse, sondern auch das Büro. Ich glaube, ich hätte mir keinen besseren Bürogenossen wünschen können. So manche Erkenntnisse zu den (Un)Tiefen der Migration ergaben sich durch unsere Diskussionen beim täglichen Kaffeetrinken. **Miriam Spinner** stieß im Laufe der Zeit zu uns hinzu und integrierte sich in kürzester Zeit in unsere Gruppe. Ihr Fleiß und ihr Engagement sind wirklich bemerkenswert. Natürlich gilt mein Dank auch den früheren Mitgliedern der Arbeitsgruppe, **Dr. Alexander Goertz** und **Dr. Matthias Riede**. Ich denke, wir waren ein gutes Team und legten damals gemeinsam den Grundstein für eine erfolgreiche Arbeit. An **Andreas Hippel** geht mein Dank für die Hilfe bei der Programmierung.

Ein besonderes Dankeschön geht an **Dr. Jürgen Mann**. Ich wage mir nicht auszumalen, wie weit mein Programm ohne seine Ratschläge und Unterstützung in Sachen C++ gediehen wäre. Es fasziniert mich immer wieder, wie er als wandelndes Lexikon souverän mit technischen Dingen und komplizierten Sachverhalten umgeht – es hätte mich wahrlich nicht verwundert, wenn er seine eigene Dissertation direkt in PostScript geschrieben hätte. Viele Studierende und Doktoranden konnten bereits von seiner Hilfsbereitschaft profitieren. **Eric Duveneck** war sehr hilfreich bei der Verarbeitung des Realdatensatzes, der in Kapitel 9 gezeigt wird, und stellte mir u. a. das Geschwindigkeitsmodell zur Verfügung. Von seinem sehr breiten Wissen im Bereich der Seismik konnte ich immer wieder profitieren. Mit **Steffen Bergler** verbrachte ich so manche Stunde im Hörsaal, um dort unser Wissen den Studierenden weiterzugeben. Seine stets ruhige und besonnene Art hat sich in vielen Diskussionen ausgezahlt. **Zeno Heilmann** gilt mein Dank für die Zusammenarbeit bei der Verarbeitung der Realdaten (Kapitel 9) und **Dr. German Höcht** für die wertvollen C++ Routinen, die mir so manche zusätzliche Arbeit erspart haben. An dieser Stelle sei auch allen anderen (Ex-)Mitgliedern der Seismik-Arbeitsgruppe gedankt, seien es Studierende, Diplomanden oder Doktoranden – alle tragen (bzw. trugen) auf ihre Weise zum guten Arbeitsklima und damit zum Erfolg der Gruppe und des WIT Konsortiums bei.

Ich danke den Korrekturlesern meiner Arbeit, **Jürgen**, **Eric**, **Christoph**, **Steffen** und **Alexander**, die meine Kapitel auf inhaltliche, grammatikalische und orthographische Fehler durchforstet haben und zahlreiche Verbesserungsvorschläge einbrachten. Ferner danke ich **Jan Würfel** und **Marcus Diem**, die Auszüge meiner Arbeit korrekturgelesen haben – ihre Kommentare aus Sicht eines Nicht-Geophysikers haben sehr dazu beigetragen, manche Dinge klarer und einfacher darzustellen und somit für eine bessere Verständlichkeit des Textes zu sorgen. Ein besonderes Dankeschön geht an **Maren Böse**, die ebenfalls Teile meiner Arbeit korrekturgelesen hat und mich auf den täglichen Fahrten zum und vom Institut ertragen musste ;–) Ich hoffe, ihr haben die gemeinsamen Rad- und Straßenbahnfahrten genauso viel Spaß gemacht wie mir. Ich werde diese ganz sicher vermissen.

Der HotRock EWK Offenbach/Pfalz GmbH danke ich für die Erlaubnis, die Realdaten (Kapitel 9) veröffentlichen zu dürfen. Insbesondere geht der Dank an **Dr. Horst Kreuter** und **Prof. Norman Harthill** für ihre Hilfsbereitschaft und ihr Interesse, das sie an meinen Arbeiten gezeigt haben. Die Deutsche Montan Technologie GmbH, die die Realdaten im Feld akquirierte und die das Preprocessing übernahm, gab bereitwillig Einblick in und Auskunft über die Verarbeitung der Daten – hierfür sei ebenfalls gedankt.

I thank the sponsors of the Wave Inversion Technoloy (WIT) consortium for their ongoing support and the WIT working groups for their cooperation. It is worthwhile to mention especially the contributions of **Dr. Claudia Vanelle**.

I am grateful to **Maren Scheidhauer** and **Prof. Anne Trehu** for providing the marine dataset that is shown in Chapter 6.

An **Claudia Payne** geht ein besonders herzliches Dankeschön. Sie behielt auch in stürmischen Zeiten immer den Überblick und half mir, so manches bürokratische Hindernis zu überwinden. Ihre täglichen Besuche in meinem Büro werde ich vermissen. Was wäre die Arbeitsgruppe ohne ihr Engagement?

Ohne die Unterstützung von **Petra Knopf** und **Thomas Nadolny** wäre der effiziente Umgang mit der am Geophysikalischen Institut betriebenen Hardware nicht möglich gewesen. Durch ihre Arbeit hinter den Kulissen wurde nicht nur meine Dissertation, sondern auch die von vielen anderen erst ermöglicht. Petra danke ich ferner für ihr Vertrauen, mir in ihrer Abwesenheit die Obhut über das Rechnernetzwerk des Instituts zu überlassen.

Prof. Helmut Wilhelm danke ich dafür, dass er mir immer wieder Hintergrundinformationen zum Studium der Geophysik im Allgemeinen und an der Universität Karlsruhe im Speziellen zukommen ließ. Er hat mir sehr geholfen, meine Aufgabe als Studienberater zu erfüllen. Ebenso danke ich allen anderen Mitgliedern des Geophysikalischen Instituts, die hier nicht explizit genannt wurden (seien sie aus der Verwaltung, dem technischen oder wissenschaftlichen Dienst). Sie sind nicht in Vergessenheit geraten.

All meinen Freunden in nah und fern ein herzliches „Danke" für gemeinsame spaßige Zeiten und viele Emails: **Maren**, **Christoph**, **Eric**, **Jürgen**, **Steffen**, **Jan**, **Heidi**, **Marcus**, **Max**, **Simone**, **Ted**, **Ulrich**, **Suzan**, **Teresa**, **Martin**, **Dagmar**, **Carmen**, **Miriam**, **Paola**, **Alex**, **Matthias**, **Anne**, **Tina**, **Patrick**, **Hanna**, **Markus**, **Jochen**, **Ingo**, **Seppl**, **Philipp**, **Silke**, **Sven**, und und und. . . , besonders auch für die Unterstützung in Zeiten, die nicht so einfach für mich waren.

Claudine Groß hat mich eine Zeit lang auf meinem Lebensweg begleitet und ihn sehr bereichert. Sie hat mir abseits der Diplom- und Doktorarbeit immer wieder die Dinge gezeigt, die das Leben lebenswert machen. Ich würde ihr gern so viele Dinge sagen, aber Worte sind manchmal einfach fehl am Platz. Ich hoffe, sie versteht mich auch so – deswegen hier einfach von ganzem Herzen ein schlichtes „Danke!".

Das Beste hebt man sich bekanntlich immer bis zum Schluss auf: Mein ganz besonderer Dank gilt meinen Eltern, **Elisabeth** und **Fritz Hertweck**. Nicht nur, dass sie mir das Studium der Geophysik erst ermöglichten, sie unterstützten mich auch sonst in jeglicher Hinsicht mit Rat und Tat. Ihnen ist diese Dissertation gewidmet.

Lebenslauf

Persönliche Daten

Name:	Thomas Hertweck
Geburtsdatum:	07. Mai 1974
Nationalität:	deutsch
Geburtsort:	Baden-Baden

Schulausbildung

1980 - 1984	Theodor-Heuss-Grundschule Baden-Baden
1984 - 1993	Markgraf-Ludwig-Gymnasium Baden-Baden
17.05.1993	Allgemeine Hochschulreife

Hochschulausbildung

1993 - 1995 & 1996 - 2000	Studium der Geophysik an der Universität Karlsruhe (TH)
17.05.2000	Diplom
seit Aug. 2000	Doktorand an der Fakultät für Physik der Universität Karlsruhe (TH)